T0213295

SpringerBriefs in Applied Sciences and Technology

Continuum Mechanics

Series Editors

Holm Altenbach, Institut für Mechanik, Lehrstuhl für Technische Mechanik, Otto von Guericke University Magdeburg, Magdeburg, Sachsen-Anhalt, Germany

Andreas Öchsner, Faculty of Mechanical Engineering, Esslingen University of Applied Sciences, Esslingen am Neckar, Germany

These SpringerBriefs publish concise summaries of cutting-edge research and practical applications on any subject of Continuum Mechanics and Generalized Continua, including the theory of elasticity, heat conduction, thermodynamics, electromagnetic continua, as well as applied mathematics.

SpringerBriefs in Continuum Mechanics are devoted to the publication of fundamentals and applications, presenting concise summaries of cutting-edge research and practical applications across a wide spectrum of fields. Featuring compact volumes of 50 to 125 pages, the series covers a range of content from professional to academic.

More information about this subseries at http://www.springer.com/series/10528

Andreas Öchsner

Partial Differential Equations of Classical Structural Members

A Consistent Approach

 Springer

Andreas Öchsner
Faculty of Mechanical Engineering
Esslingen University of Applied Sciences
Esslingen, Baden-Württemberg, Germany

ISSN 2191-530X   ISSN 2191-5318 (electronic)
SpringerBriefs in Applied Sciences and Technology
ISSN 2625-1329   ISSN 2625-1337 (electronic)
SpringerBriefs in Continuum Mechanics
ISBN 978-3-030-35310-0   ISBN 978-3-030-35311-7   (eBook)
https://doi.org/10.1007/978-3-030-35311-7

This Springer imprint is published by the registered company Springer Nature Switzerland AG
The registered company address is: Gewerbestrasse 11, 6330 Cham, Switzerland

Preface

Partial differential equations (PDEs) form the basis to mathematically describe the mechanical behavior of all classical structural members known in engineering mechanics. Nevertheless, there are concerning trends in some places of tertiary education to no more derive and work with these equations. This may lead to a serious lack of knowledge and may affect the reliable design of engineering structures and processes.

The derivation and understanding of PDEs relies heavily on the fundamental knowledge of the first years of engineering education, i.e., higher mathematics, physics, materials science, applied mechanics, design, and programming skills. Thus, it is definitely a challenging topic for prospective engineers.

This volume in the SpringerBriefs series should provide a compact overview on the classical PDEs of structural members and to provide a formal way to uniformly describe these equations in a similar way. All derivations in the following chapters follow a common approach: the three fundamental equations of continuum mechanics, i.e., the kinematics equation, the constitutive equation, and the equilibrium equation, are combined to construct the partial differential equations.

The structural members covered in this book are rods, thin and thick beams, plane elasticity members, thin and thick plates, and three-dimensional solids. The last chapter gives a brief introduction to the topic of transient analysis.

Esslingen, Germany   Andreas Öchsner
October 2019

Contents

Chapter 1
Introduction to Structural Modeling

Abstract The first chapter classifies the content as well as the focus of this textbook. In engineering practice, the description of processes is centered around partial differential equations, and all the classical approximation methods such as the finite element method, the finite difference method, the finite volume method, and the boundary element method offer different ways of solving these equations.

Engineers describe physical phenomena and processes typically by equations, particularly by partial differential equations [3, 4, 9]. In this context, the derivation and the solution of these differential equations (see Fig. 1.1) is the task of engineers, obviously requiring fundamental knowledge from physics and engineering mathematics. The classical engineering methods for providing approximate solutions of partial differential equations are the finite element method (FEM), the finite difference method (FDM), the finite volume method (FVM), and the boundary element method (BEM) [1, 2, 8].

The importance of partial differential equations is clearly represented in the following quote: 'For more than 250 years partial differential equations have been clearly the most important tool available to mankind in order to understand a large variety of phenomena, natural at first and then those originating from human activity and technological development. Mechanics, physics and their engineering applications were the first to benefit from the impact of partial differential equations on modeling and design,...' [5].

In the one-dimensional case, a physical problem can be generally described in a spatial domain Ω by the differential equation

$$\mathcal{L}\{y(x)\} = b \quad (x \in \Omega) \tag{1.1}$$

and by the conditions which are prescribed on the boundary Γ. The differential equation is also called the *strong form* or the *original statement* of the problem. The expression 'strong form' comes from the fact that the differential equation describes exactly each point x in the domain of the problem. The operator $\mathcal{L}\{\ldots\}$ in Eq. (1.1) is an arbitrary differential operator which can take, for example, the following forms:

© The Author(s), under exclusive license to Springer Nature Switzerland AG 2020
A. Öchsner, *Partial Differential Equations of Classical Structural Members*,
SpringerBriefs in Continuum Mechanics,
https://doi.org/10.1007/978-3-030-35311-7_1

1

Fig. 1.1 Modeling based on
partial differential equations

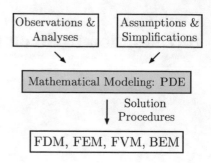

$$\mathcal{L}\{\ldots\} = \frac{\mathrm{d}^2}{\mathrm{d}x^2}\{\ldots\}, \tag{1.2}$$

$$\mathcal{L}\{\ldots\} = \frac{\mathrm{d}^4}{\mathrm{d}x^4}\{\ldots\}, \tag{1.3}$$

$$\mathcal{L}\{\ldots\} = \frac{\mathrm{d}^4}{\mathrm{d}x^4}\{\ldots\} + \frac{\mathrm{d}}{\mathrm{d}x}\{\ldots\} + \{\ldots\}. \tag{1.4}$$

Furthermore, variable b in Eq. (1.1) is a given function, and in the case of $b = 0$, the equation reduces to the *homogeneous differential equation*: $\mathcal{L}\{y(x)\} = 0$. More specific expressions of Eqs. (1.3) till (1.4) can take the following form [6]:

$$a\frac{\mathrm{d}^2 y(x)}{\mathrm{d}x^2} = b, \tag{1.5}$$

$$a\frac{\mathrm{d}^4 y(x)}{\mathrm{d}x^4} = b, \tag{1.6}$$

and will be used to describe the behavior of rods and beams in the following sections.

Let us highlight at the end of this section that the derivations in the following chapters follow a common approach, see Fig. 1.2 [7].

A combination of the kinematics equation (i.e., the relation between the strains and displacements) with the constitutive equation (i.e., the relation between the stresses and strains) and the equilibrium equation (i.e., the equilibrium between the internal reactions and the external loads) results in a partial differential equation. Limited to simple cases, analytical solutions can be derived and are exact under the given assumptions. For more complex problems, engineers rely on the classical numerical methods as already mentioned in this chapter.

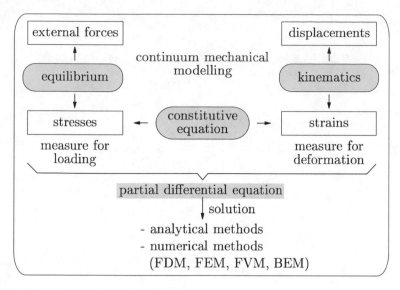

Fig. 1.2 Continuum mechanical modelling

References

1. Bathe K-J (1996) Finite element procedures. Prentice-Hall, Upper Saddle River
2. Cook RD, Malkus DS, Plesha ME, Witt RJ (2002) Concepts and applications of finite element analysis. Wiley, New York
3. Debnath L (2012) Nonlinear partial differential equations for scientists and engineers. Springer, New York
4. Formaggia L, Saleri F, Veneziani A (2012) Solving numerical PDEs: problems, applications, exercises. Springer, Milan
5. Glowinski R, Neittaanmki P (eds) (2008) Partial differential equations: modelling and numerical simulation. Springer, Dordrecht
6. Öchsner A (2014) Elasto-plasticity of frame structure elements: modeling and simulation of rods and beams. Springer, Berlin
7. Öchsner A (2016) Continuum damage and fracture mechanics. Springer, Singapore
8. Öchsner A (2018) A project-based introduction to computational statics. Springer, Cham
9. Salsa S (2008) Partial differential equations in action: from modelling to theory. Springer, Milano

Chapter 2
Rods or Bars

Abstract This chapter covers the continuum mechanical description of rod/bar members. Based on the three basic equations of continuum mechanics, i.e., the kinematics relationship, the constitutive law and the equilibrium equation, the partial differential equation, which describes the physical problem, is derived.

2.1 Introduction

A rod is defined as a prismatic body whose axial dimension is much larger than its transverse dimensions. This structural member is only loaded in the direction of the main body axes, see Fig. 2.1a. As a result of this loading, the deformation occurs only along its main axis.

The following derivations are restricted to some simplifications:

- only applying to straight rods,
- displacements are (infinitesimally) small,
- strains are (infinitesimally) small, and
- material is linear-elastic (homogeneous and isotropic).

It should be noted here that the alternatively nomenclature 'bar' is also found in scientific literature to describe a rod member. Details on the continuum mechanical description of rods can be found in [1, 2] and the basic equations are derived in detail in the following sections.

2.2 Kinematics

To derive the strain-displacement relation (kinematics relation), an axially loaded rod is considered as shown in Fig. 2.2. The length of the member is equal to L and the constant axial tensile stiffness is equal to EA. The load is either given as a single force F_x and/or as a distributed load $p_x(x)$.

This distributed load has the unit of force per unit length. In the case of a body force f_x (unit: force per unit volume), the distributed load takes the form $p_x(x) =$

A. Öchsner, *Partial Differential Equations of Classical Structural Members*,
SpringerBriefs in Continuum Mechanics,
https://doi.org/10.1007/978-3-030-35311-7_2

Fig. 2.1 Schematics representation of a continuum rod

Fig. 2.2 General
configuration of an axially
loaded rod: **a** geometry and
material property; **b**
prescribed loads

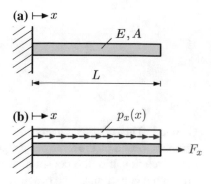

Fig. 2.3 Elongation of a
differential element of length
dx: **a** undeformed
configuration; **b** deformed
configuration

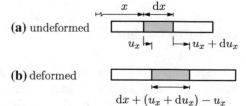

$f_x(x)A(x)$ where A is the cross-sectional area of the rod. A typical example for a body force would be the dead weight, i.e. the mass under the influence of gravity. In the case of a traction force t_x (unit: force per unit area), the distributed load can be written as $p_x(x) = t_x(x)U(x)$ where $U(x)$ is the perimeter of the cross section. Typical examples are frictional resistance, viscous drag and surface shear.

Let us now consider a differential element dx of such a rod as shown in Fig. 2.3. Under an acting load, this element deforms as indicated in Fig. 2.3b where the initial point at the position x is displaced by u_x and the end point at the position $x + dx$ is displaced by $u_x + du_x$. Thus, the differential element which has a length of dx in the unloaded state elongates to a length of $dx + (u_x + du_x) - u_x$.

The engineering strain, i.e., the increase in length related to the original length, can be expressed as

$$\varepsilon_x = \frac{(dx + (u_x + du_x) - u_x) - (dx)}{dx}, \tag{2.1}$$

or finally as:

$$\varepsilon_x(x) = \frac{du_x(x)}{dx}. \tag{2.2}$$

Fig. 2.4 Axially loaded rod:
a strain and b stress
distribution

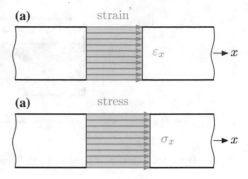

Fig. 2.4 Axially loaded rod: a strain and b stress distribution

The last equation is often expressed in a less mathematical way (non-differential) as $\varepsilon_x = \frac{\Delta L}{L}$ where ΔL is the change in length of the entire rod element.

2.3 Constitution

The constitutive equation, i.e., the relation between the stress σ_x and the strain ε_x, is given in its simplest form as Hooke's law

$$\sigma_x(x) = E\varepsilon_x(x), \tag{2.3}$$

where the Young's modulus E is in the case of linear elasticity a material constant. For the considered rod element, the normal stress and strain is constant over the cross section as shown in Fig. 2.4.

2.4 Equilibrium

The equilibrium equation between the external forces and internal reactions can be derived for a differential element of length dx as shown in Fig. 2.5. It is assumed for simplicity that the distributed load p_x and the cross-sectional area A are constant in this figure. The internal reactions N_x are drawn in their positive directions, i.e., at the left-hand face in the negative and at the right-hand face in the positive x-direction.

The force equilibrium in the x-direction for a static configuration requires that

$$- N_x(x) + p_x dx + N_x(x + dx) = 0 \tag{2.4}$$

holds. A first-order Taylor's series expansion of the normal force $N_x(x + dx)$ around point x, i.e.

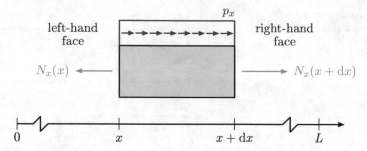

Fig. 2.5 Differential element of a rod with internal reactions and constant external distributed load

Table 2.1 Fundamental governing equations of a rod for deformation along the x-axis

Expression	Equation
Kinematics	$\varepsilon_x(x) = \dfrac{\mathrm{d}u_x(x)}{\mathrm{d}x}$
Equilibrium	$\dfrac{\mathrm{d}N_x(x)}{\mathrm{d}x} = -p_x(x)$
Constitution	$\sigma_x(x) = E\varepsilon_x(x)$

$$N_x(x + \mathrm{d}x) \approx N_x(x) + \left.\frac{\mathrm{d}N_x}{\mathrm{d}x}\right|_x \mathrm{d}x \,, \tag{2.5}$$

allows to finally express Eq. (2.4) as:

$$\frac{\mathrm{d}N_x(x)}{\mathrm{d}x} = -p_x(x) \,. \tag{2.6}$$

The three fundamental equations to describe the behavior of a rod element are summarized in Table 2.1.

A slightly different derivation of the equilibrium equation is obtained as follows: Eq. (2.4) can be expressed based on the normal stresses as:

$$-\sigma_x(x)A + p_x\mathrm{d}x + \sigma_x(x + \mathrm{d}x)A = 0 \,. \tag{2.7}$$

A first-order Taylor's series expansion of the stress $\sigma_x(x + \mathrm{d}x)$ around point x, i.e.

$$\sigma_x(x + \mathrm{d}x) \approx \sigma_x(x) + \left.\frac{\mathrm{d}\sigma_x}{\mathrm{d}x}\right|_x \mathrm{d}x \,, \tag{2.8}$$

allows to finally express Eq. (2.7) as:

$$\frac{\mathrm{d}\sigma_x(x)}{\mathrm{d}x} + \frac{p_x(x)}{A} = 0 \,. \tag{2.9}$$

The last equation with $\sigma_x = \frac{N_x}{A}$ immediately gives Eq. (2.6).

2.5 Differential Equation

To derive the governing partial differential equation, the three fundamental equations given in Table 2.1 must be combined. Introducing the kinematics relation (2.2) into Hooke's law (2.3) gives:

$$\sigma_x(x) = E\frac{du_x}{dx}. \tag{2.10}$$

Considering in the last equation that a normal stress is defined as an acting force N_x over a cross-sectional area A:

$$\frac{N_x}{A} = E\frac{du_x}{dx}. \tag{2.11}$$

The last equation can be differentiated with respect to the x-coordinate to give:

$$\frac{dN_x}{dx} = \frac{d}{dx}\left(EA\frac{du_x}{dx}\right), \tag{2.12}$$

where the derivative of the normal force can be replaced by the equilibrium equation (2.6) to obtain in the general case:

$$\frac{d}{dx}\left(E(x)A(x)\frac{du_x(x)}{dx}\right) = -p_x(x). \tag{2.13}$$

The general case in the formulation with the internal normal force distribution reads:

$$E(x)A(x)\frac{du_x(x)}{dx} = N_x(x). \tag{2.14}$$

Thus, to obtain the displacement field $u_x(x)$, one may start from Eq. (2.13) or from Eq. (2.14). The first approach requires to state the distribution of the distributed load $p_x(x)$ while for the second approach one requires the internal normal force distribution $N_x(x)$.

If the axial tensile stiffness EA is constant, the formulation (2.13) can be simplified to:

$$EA\frac{d^2u_x(x)}{dx^2} = -p_x(x). \tag{2.15}$$

Some common formulations of the governing partial differential equation are collected in Table 2.2. It should be noted here that some of the different cases given in Table 2.2 can be combined. The last case in Table 2.2 refers to the case of elastic embedding of a rod where the embedding modulus k has the unit of force per unit

Table 2.2 Different formulations of the partial differential equation for a rod (x-axis: right facing)

Configuration	Partial differential equation
E, A	$EA\dfrac{d^2u_x}{dx^2} = 0$
$E(x), A(x)$	$\dfrac{d}{dx}\left(E(x)A(x)\dfrac{du_x}{dx}\right) = 0$
$p_x(x)$	$EA\dfrac{d^2u_x}{dx^2} = -p_x(x)$
$k(x)$	$EA\dfrac{d^2u_x}{dx^2} = k(x)u_x$

Table 2.3 Different formulations of the basic equations for a rod (x-axis along the principal rod axis). E: Young's modulus; A: cross-sectional area; p_x: length-specific distributed normal load; $\mathcal{L}_1 = \frac{d(\ldots)}{dx}$: first-order derivative; b: volume-specific distributed normal load

Specific formulation	General formulation
Kinematics	
$\varepsilon_x(x) = \dfrac{du_x(x)}{dx}$	$\varepsilon_x(x) = \mathcal{L}_1(u_x(x))$
Constitution	
$\sigma_x(x) = E\varepsilon_x(x)$	$\sigma_x(x) = C\varepsilon_x(x)$
Equilibrium	
$\dfrac{d\sigma_x(x)}{dx} + \dfrac{p_x(x)}{A} = 0$	$\mathcal{L}_1^{\mathrm{T}}(\sigma_x(x)) + b = 0$
PDE (A = const.)	
$\dfrac{d}{dx}\left(E(x)\dfrac{du_x}{dx}\right) + \dfrac{p_x(x)}{A} = 0$	$\mathcal{L}_1^{\mathrm{T}}(C\mathcal{L}_1(u_x(x))) + b = 0$
	or
	$\mathcal{L}_1^{\mathrm{T}}(EA\mathcal{L}_1(u_x(x))) + p_x = 0$

area. Analytical solutions for different loading and support conditions can be found, for example, in [3].

If we replace the common formulation of the first order derivative, i.e. $\frac{d(\ldots)}{dx}$, by a formal operator symbol, i.e. $\mathcal{L}_1(\ldots)$, the basic equations can be stated in a more formal way as given in Table 2.3. Such a formulation is advantageous in the two- and three-dimensional cases. It should be noted here that the transposed ('T'), i.e., $\mathcal{L}_1^{\mathrm{T}} = \left(\frac{d(\ldots)}{dx}\right)^{\mathrm{T}}$, is only used to show later similar structures of the equations in the two- and three-dimensional case.

References

1. Gross D, Hauger W, Schrder J, Wall WA, Bonet J (2011) Engineering mechanics 2: mechanics of materials. Springer, Berlin
2. Hartsuijker C, Welleman JW (2007) Engineering mechanics volume 2: stresses, strains, displacements. Springer, Dordrecht
3. Öchsner A (2014) Elasto-plasticity of frame structure elements: modeling and simulation of rods and beams. Springer, Berlin

Chapter 3
Euler–Bernoulli Beams

Abstract This chapter covers the continuum mechanical description of thin beam members. Based on the three basic equations of continuum mechanics, i.e., the kinematics relationship, the constitutive law and the equilibrium equation, the partial differential equation, which describes the physical problem, is derived.

3.1 Introduction

A beam is defined as a long prismatic body as schematically shown in Fig. 3.1a. The following derivations are restricted to some simplifications:

- only applying to straight beams,
- no elongation along the x-axis,
- no torsion around the x-axis,
- deformations in a single plane, i.e. symmetrical bending,
- small deformations, and
- simple cross sections.

The external loads, which are considered within this chapter, are the single forces F_z, single moments M_y, distributed loads $q_z(x)$, and distributed moments $m_y(x)$. These loads have in common that their line of action (force) or the direction of the momentum vector are orthogonal to the center line of the beam and cause its bending. This is a different type of deformation compared to the rod element from Chap. 2, see Table 3.1. It should be noted here that these basic types of deformation can be superposed to account for more complex loading conditions [2].

The classic theories of beam bending distinguish between shear-rigid and shear-flexible models. The shear rigid-beam, also called the Bernoulli beam,[1] neglects the shear deformation from the shear forces. This theory implies that a cross-sectional plane which was perpendicular to the beam axis before the deformation remains in

---

[1]More precisely, this beam is known as the Euler–Bernoulli beam. A historical analysis of the development of the classical beam theory and the contribution of different scientists can be found in [7].

A. Öchsner, *Partial Differential Equations of Classical Structural Members*,
SpringerBriefs in Continuum Mechanics,
https://doi.org/10.1007/978-3-030-35311-7_3

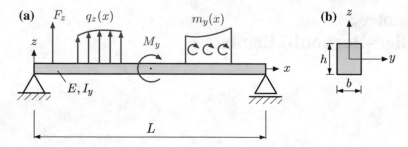

Fig. 3.1 General configuration for beam problems: **a** example of boundary conditions and external loads; **b** cross-sectional area

Table 3.1 Differentiation between rod and beam element; center line parallel to the x-axis

	Rod	Beam
Force	Along the rod axis	Perpendicular to the beam axis
Unknown	Displacement along rod axis	Displacement perpendicular and rotation perpendicular to the beam axis
	u_x	u_z, φ_y

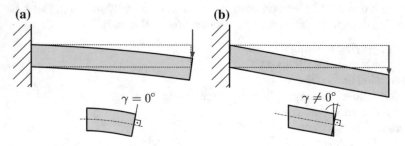

Fig. 3.2 Different deformation modes of a bending beam: **a** shear-rigid; **b** shear-flexible. Adapted from [6]

the deformed state perpendicular to the beam axis, see Fig. 3.2a. Furthermore, it is assumed that a cross-sectional plane stays plane and unwarped in the deformed state. These two assumptions are also known as Bernoulli's hypothesis. Altogether one imagines that cross-sectional planes are rigidly fixed to the center line of the beam[2] so that a change of the center line affects the entire deformation. Consequently, it is also assumed that the geometric dimensions[3] of the cross-sectional planes do not change.

---

[2]More precisely, this is the neutral fibre or the bending line.
[3]Consequently, the width b and the height h of a, for example, rectangular cross section remain the same, see Fig. 3.1b.

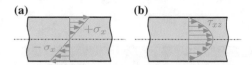

Fig. 3.3 Different stress distributions of a beam with rectangular cross section and linear-elastic material behavior: **a** normal stress and **b** shear stress

Table 3.2 Analogies between the beam and plate theories

	Beam theory	Plate theory
Dimensionality	1D	2D
Shear-rigid	Bernoulli beam	Kirchhoff plate
Shear-flexible	Timoshenko beam	Reissner–Mindlin plate

In the case of a shear-flexible beam, also called the Timoshenko beam, the shear deformation is considered in addition to the bending deformation and cross-sectional planes are rotated by an angle γ compared to the perpendicular line, see Fig. 3.2b. For beams for which the length is 10–20 times larger than a characteristic dimension of the cross section, the shear fraction is usually disregarded in the first approximation. The different load types, meaning pure bending moment loading or shear due to shear force, lead to different stress fractions in a beam. In the case of a Bernoulli beam, deformation occurs solely through normal forces, which are linearly distributed over the cross section. Consequently, a tension—alternatively a compression maximum on the bottom face—maximum on the top face occurs, see Fig. 3.3a. In the case of symmetric cross sections, the zero crossing[4] occurs in the middle of the cross section. The shear stress distribution for a rectangular cross section is parabolic at which the maximum occurs at the neutral axis and is zero at both the top and bottom surface, see Fig. 3.3b. This shear stress distribution can be calculated for the Bernoulli beam but is not considered for the derivation of the deformation.

Finally, it needs to be noted that the one-dimensional beam theories have corresponding counterparts in two-dimensional space, see Table 3.2. In plate theories, the Bernoulli beam corresponds to the shear-rigid plate and the Timoshenko beam corresponds to the shear-flexible Reissner–Mindlin plate, [1, 4, 11].

Further details regarding the beam theory and the corresponding basic definitions and assumptions can be found in references [3, 5, 8, 10]. In the following sections, only the Bernoulli beam is considered. Consideration of the shear part takes place in Chap. 4.

3.2 Kinematics

For the derivation of the kinematics relation, a beam with length L is under constant moment loading $M_y(x) = \text{const.}$, meaning under *pure* bending, is considered, see

---

[4]The sum of all points with $\sigma = 0$ along the beam axis is called the neutral fiber.

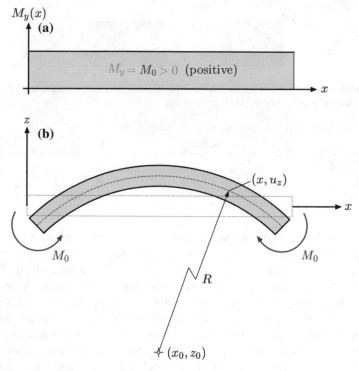

Fig. 3.4 Beam under pure bending in the x-z plane: **a** moment distribution; **b** deformed beam. Note that the deformation is exaggerated for better illustration. For the deformations considered in this chapter the following applies: $R \gg L$

Fig. 3.4. One can see that both external single moments at the left- and right-hand boundary lead to a positive bending moment distribution M_y within the beam. The vertical position of a point with respect to the center line of the beam *without action* of an external load is described through the z-coordinate. The vertical *displacement* of a point on the center line of the beam, meaning for a point with $z = 0$, under action of the external load is indicated with u_z. The deformed center line is represented by the sum of these points with $z = 0$ and is referred to as the bending line $u_z(x)$.

In the case of a deformation in the x-z plane, it is important to precisely distinguish between the positive orientation of the internal reactions, the positive rotational angle, and the slope see Fig. 3.5. The internal reactions at a right-hand boundary are directed in the positive directions of the coordinate axes. Thus, a positive moment at a right-hand boundary is clockwise oriented (as the positive rotational angle), see Fig. 3.5. However, the slope is negative, see Fig. 3.5. This difference requires some careful derivations of the corresponding equations.

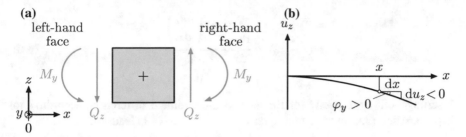

Fig. 3.5 Positive definition of **a** internal reactions and **b** rotation (but negative slope)

Only the center line of the deformed beam is considered in the following. Through the relation for an arbitrary point (x, u_z) on a circle with radius R around the center point (x_0, z_0), meaning

$$(x - x_0)^2 + (u_z(x) - z_0)^2 = R^2, \tag{3.1}$$

one obtains through differentiation with respect to the x-coordinate

$$2(x - x_0) + 2(u_z(x) - z_0)\frac{du_z(x)}{dx} = 0, \tag{3.2}$$

alternatively after another differentiation:

$$2 + 2\frac{du_z}{dx}\frac{du_z}{dx} + 2(u_z(x) - z_0)\frac{d^2u_z}{dx^2} = 0. \tag{3.3}$$

Equation (3.3) provides the vertical distance between an arbitrary point on the center line of the beam and the center point of a circle as

$$(u_z - z_0) = -\frac{1 + \left(\dfrac{du_z}{dx}\right)^2}{\dfrac{d^2u_z}{dx^2}}, \tag{3.4}$$

while the difference between the x-coordinates results from Eq. (3.2):

$$(x - x_0) = -(u_z - z_0)\frac{du_z}{dx}. \tag{3.5}$$

If the expression according to Eq. (3.4) is used in Eq. (3.5) the following results:

$$(x - x_0) = \frac{du_z}{dx} \frac{1 + \left(\frac{du_z}{dx}\right)^2}{\frac{d^2 u_z}{dx^2}}.$$

(3.6)

Inserting both expressions for the x- and z-coordinate differences according to Eqs. (3.6) and (3.4) in the circle equation according to (3.1) leads to:

$$R^2 = (x - x_0)^2 + (u_z - z_0)^2$$

(3.7)

$$= \left(\frac{du_z}{dx}\right)^2 \frac{\left(1 + \left(\frac{du_z}{dx}\right)^2\right)^2}{\left(\frac{d^2 u_z}{dx^2}\right)^2} + \frac{\left(1 + \left(\frac{du_z}{dx}\right)^2\right)^2}{\left(\frac{d^2 u_z}{dx^2}\right)^2}$$

$$= \left(\left(\frac{d^2 u_z}{dx^2}\right)^2 + 1\right) \frac{\left(1 + \left(\frac{du_z}{dx}\right)^2\right)^2}{\left(\frac{d^2 u_z}{dx^2}\right)^2}$$

$$= \frac{\left(1 + \left(\frac{du_z}{dx}\right)^2\right)^3}{\left(\frac{d^2 u_z}{dx^2}\right)^2}.$$

(3.8)

Thus, the radius of curvature is obtained as:

$$|R| = \frac{\left(1 + \left(\frac{du_z}{dx}\right)^2\right)^{3/2}}{\left|\frac{d^2 u_z}{dx^2}\right|}.$$

(3.9)

To decide if the radius of curvature is positive or negative, let us have a look at Fig. 3.6 where a curve with its tangential and normal vectors is shown. Since the curve in this configuration is bending away from the normal vector $\boldsymbol{n}$, it holds that $\frac{d^2 u_z}{dx^2} < 0$ and the radius of curvature is obtained for a positive bending moment as:

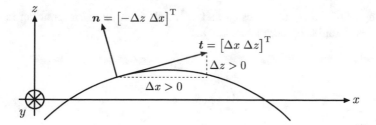

Fig. 3.6 On the definition of a negative curvature in the x-z plane

Fig. 3.7 Segment of a beam under pure bending in the x-z plane. Note that the deformation is exaggerated for better illustration

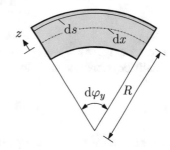

$$R = -\frac{\left(1 + \left(\dfrac{du_z}{dx}\right)^2\right)^{3/2}}{\dfrac{d^2 u_z}{dx^2}}.$$ (3.10)

Note that the expression curvature, which results as a reciprocal value from the curvature radius, $\kappa = \frac{1}{R}$, is used as well.

For small bending deflections, meaning $u_z \ll L$, $\frac{du_z}{dx} \ll 1$ results and Eq. (3.10) simplifies to:

$$R = -\frac{1}{\dfrac{d^2 u_z}{dx^2}} \quad \text{or} \quad \kappa = \frac{1}{R} = -\frac{d^2 u_z}{dx^2}.$$ (3.11)

For the determination of the strain, one refers to its general definition, meaning elongation referring to initial length. Relating to the configuration shown in Fig. 3.7, the longitudinal elongation of a fibre at distance z to the neutral fibre allows to express the strain as:

$$\varepsilon_x = \frac{ds - dx}{dx}.$$ (3.12)

The lengths of the circular arcs ds and dx result from the corresponding radii and the enclosed angles in radian measure as:

$$dx = R d\varphi_y,$$ (3.13)

$$ds = (R + z) d\varphi_y.$$ (3.14)

If these relations for the circular arcs are used in Eq. (3.12), the following results:

$$\varepsilon_x = \frac{(R + z) d\varphi_y - R d\varphi_y}{dx} = z \frac{d\varphi_y}{dx}.$$ (3.15)

From Eq. (3.13) $\frac{d\varphi_y}{dx} = \frac{1}{R}$ results and together with relation (3.11) the strain can finally be expressed as follows:

$$\varepsilon_x(x, y) = z \frac{1}{R} \overset{(3.11)}{=} -z \frac{d^2 u_z(x)}{dx^2} \overset{(3.11)}{=} z\kappa.$$ (3.16)

An alternative derivation of the kinematics relation results from consideration of Fig. 3.8. From the relation of the right-angled triangle $0'1'2'$, this means[5] $\sin \varphi_y = \frac{u_x}{z}$, the following relation results for small angles ($\sin \varphi_y \approx \varphi_y$):

$$u_x = +z\varphi_y.$$ (3.17)

Furthermore, it holds that the rotation angle of the slope equals the center line for small angles:

$$\tan \varphi_y = \frac{-d u_z(x)}{dx} \approx \varphi_y.$$ (3.18)

If Eqs. (3.18) and (3.17) are combined, the following results:

$$u_x = -z \frac{d u_z(x)}{dx}.$$ (3.19)

The last relation equals $(ds - dx)$ in Eq. (3.12) and differentiation with respect to the x-coordinate leads directly to Eq. (3.16).

---

[5]Note that according to the assumptions of the Bernoulli beam the lengths 01 and $0'1'$ remain unchanged.

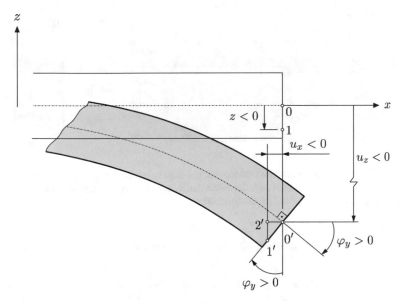

Fig. 3.8 Alternative configuration for the derivation of the kinematics relation. Note that the deformation is exaggerated for better illustration

3.3 Constitution

The one-dimensional Hooke's law according to Eq. (2.3) can also be assumed in the case of the bending beam, since, according to the requirement, only normal stresses are regarded in this section:

$$\sigma_x = E\varepsilon_x .\tag{3.20}$$

Through the kinematics relation according to Eq. (3.16), the stress results as a function of deflection to:

$$\sigma_x(x, z) = -Ez\frac{\mathrm{d}^2u_z(x)}{\mathrm{d}x^2}.\tag{3.21}$$

The stress distribution shown in Fig. 3.9a generates the internal moment, which acts in this cross section. To calculate this internal moment, the stress is multiplied by a surface element, so that the resulting force is obtained. Multiplication with the corresponding lever arm then gives the internal moment. Since the stress is linearly distributed over the height, the evaluation is done for an infinitesimally small surface element:

$$\mathrm{d}M_y = (+z)(+\sigma_x)\mathrm{d}A = z\sigma_x\mathrm{d}A .\tag{3.22}$$

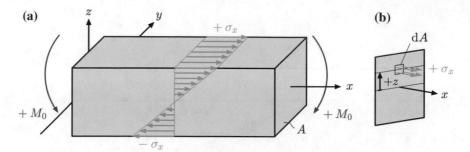

Fig. 3.9 a Schematic representation of the normal stress distribution $\sigma_x = \sigma_x(z)$ of a bending beam; **b** Definition and position of an infinitesimal surface element for the derivation of the resulting moment action due to the normal stress distribution

Therefore, the entire moment results via integration over the entire surface in:

$$M_y = \int_A z\sigma_x \mathrm{d}A \overset{(3.21)}{=} -\int_A zEz\frac{\mathrm{d}^2 u_z(x)}{\mathrm{d}x^2}\mathrm{d}A . \qquad (3.23)$$

Assuming that the Young's modulus is constant, the internal moment around the y-axis results in:

$$M_y = -E\frac{\mathrm{d}^2 u_z}{\mathrm{d}x^2}\underbrace{\int_A z^2 \mathrm{d}A}_{I_y} = \frac{I_y\sigma_z}{z}. \qquad (3.24)$$

The integral in Eq. (3.24) is the so-called axial second moment of area or axial surface moment of 2nd order in the SI unit m^4. This factor is only dependent on the geometry of the cross section and is also a measure of the stiffness of a plane cross section against bending. The values of the axial second moment of area for simple geometric cross sections are collected in Table 3.3.

Consequently the internal moment can also be expressed as

$$M_y = -EI_y\frac{\mathrm{d}^2 u_z}{\mathrm{d}x^2} \overset{(3.11)}{=} \frac{EI_y}{R} = EI_y\kappa. \qquad (3.25)$$

Equation (3.25) describes the bending line $u_z(x)$ as a function of the bending moment and is therefore also referred to as the bending line-moment relation. The product EI_y in Eq. (3.25) is also called the bending stiffness. If the result from Eq. (3.25) is used in the relation for the bending stress according to Eq. (3.21), the distribution of stress over the cross section results in:

$$\sigma_x(x, z) = +\frac{M_y(x)}{I_y}z(x) . \qquad (3.26)$$

Table 3.3 Axial second moment of area around the y- and z-axis

Cross section	I_y	I_z
	$\dfrac{\pi D^4}{64} = \dfrac{\pi R^4}{4}$	$\dfrac{\pi D^4}{64} = \dfrac{\pi R^4}{4}$
	$\dfrac{\pi b a^3}{4}$	$\dfrac{\pi a b^3}{4}$
	$\dfrac{a^4}{12}$	$\dfrac{a^4}{12}$
	$\dfrac{b h^3}{12}$	$\dfrac{h b^3}{12}$
	$\dfrac{b h^3}{36}$	$\dfrac{h b^3}{36}$
	$\dfrac{b h^3}{36}$	$\dfrac{b h^3}{48}$

The plus sign in Eq. (3.26) causes that a positive bending moment (see Fig. 3.4) leads to a tensile stress in the upper beam half (meaning for $z > 0$). The corresponding equations for a deformation in the x-y plane can be found in [9].

In the case of plane bending with $M_y(x) \neq$ const., the bending line can be approximated in each case locally through a circle of curvature, see Fig. 3.10. Therefore,

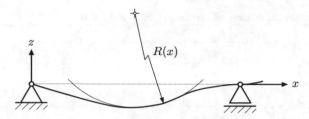

Fig. 3.10 Deformation of a beam in the x-z plane with $M_y(x) \neq$ const.

the result for *pure* bending according to Eq. (3.25) can be transferred to the case of plane bending as:

$$- E I_y \frac{\mathrm{d}^2 u_z(x)}{\mathrm{d}x^2} = M_y(x) . \tag{3.27}$$

Let us note at the end of this section that Hooke's law in the form of Eq. (3.20) is not so easy to apply[6] in the case of beams since the stress and strain is linearly changing over the height of the cross section, see Eq. (3.26) and Fig. 3.9. Thus, it might be easier to apply a so-called stress resultant or generalized stress, i.e. a simplified representation of the normal stress state[7] based on the acting bending moment:

$$M_y(x) = \iint z \sigma_x(x, z) \, \mathrm{d}A , \tag{3.28}$$

which was already introduced in Eq. (3.22). Using in addition the curvature[8] $\kappa = \kappa(x)$ (see Eq. (3.16)) instead of the strain $\varepsilon_x = \varepsilon_x(x, z)$, the constitutive equation can be easier expressed as shown in Fig. 3.11. The variables M_y and κ have both the advantage that they are constant for any location x of the beam.

3.4   Equilibrium

The equilibrium conditions are derived from an infinitesimal beam element of length $\mathrm{d}x$, which is loaded by a constant distributed load q_z, see Fig. 3.12. The internal reactions are drawn on both cut faces, i.e. at location x and $x + \mathrm{d}x$. One can see that a positive shear force is oriented in the positive z-direction at the right-hand face[9]

---

[6]However, this formulation works well in the case of rod elements since stress and strain are constant over the cross section, i.e. $\sigma_x = \sigma_x(x)$ and $\varepsilon_x = \varepsilon_x(x)$, see Fig. 2.4.

[7]A similar stress resultant can be stated for the shear stress based on the shear force: $Q_z(x) = \iint \tau_{xz}(x, z) \, \mathrm{d}A$.

[8]The curvature is then called a generalized strain.

[9]A positive cut face is defined by the surface normal on the cut plane which has the same orientation as the positive x-axis. It should be regarded that the surface normal is always directed outward.

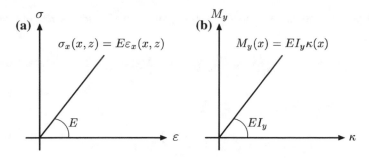

Fig. 3.11 Formulation of the constitutive law based on **a** stress and **b** stress resultant

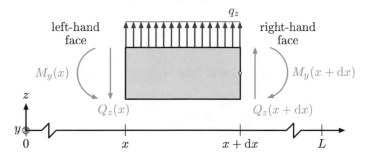

Fig. 3.12 Infinitesimal beam element in the x-z plane with internal reactions and constant distributed load

and that a positive bending moment has the same rotational direction as the positive y-axis (right-hand grip rule[10]). The orientation of shear force and bending moment is reversed at the left-hand face in order to cancel in sum the effect of the internal reactions at both faces. This convention for the direction of the internal directions is maintained in the following. Furthermore, it can be derived from Fig. 3.12 that an upwards directed *external* force or alternatively a mathematically positive oriented *external* moment at the right-hand face leads to a positive shear force or alternatively a positive internal moment. In a corresponding way, it results that a downwards directed *external* force or alternatively a mathematically negative oriented *external* moment at the left-hand face leads to a positive shear force or alternatively a positive internal moment.

The equilibrium condition will be determined in the following for the vertical forces. Assuming that forces in the direction of the positive z-axis are considered positive, the following results:

$$- Q_z(x) + Q_z(x + \mathrm{d}x) + q_z \mathrm{d}x = 0. \tag{3.29}$$

---

[10]If the axis is grasped with the right hand in a way so that the spread out thumb points in the direction of the positive axis, the bent fingers then show the direction of the positive rotational direction.

If the shear force on the right-hand face is expanded in a Taylor's series of first order, meaning

$$Q(x + \mathrm{d}x) \approx Q(x) + \frac{\mathrm{d}Q(x)}{\mathrm{d}x}\mathrm{d}x\,,$$ (3.30)

Equation (3.29) results in

$$-Q(x) + Q(x) + \frac{\mathrm{d}Q(x)}{\mathrm{d}x}\mathrm{d}x + q_z\mathrm{d}x = 0\,,$$ (3.31)

or alternatively after simplification finally to:

$$\frac{\mathrm{d}Q_z(x)}{\mathrm{d}x} = -q_z\,.$$ (3.32)

For the special case that no distributed load is acting ($q_z = 0$), Eq. (3.32) simplifies to:

$$\frac{\mathrm{d}Q(x)}{\mathrm{d}x} = 0\,.$$ (3.33)

The equilibrium of moments around the reference point at $x + \mathrm{d}x$ gives:

$$M_y(x + \mathrm{d}x) - M_y(x) - Q_z(x)\mathrm{d}x + \frac{1}{2}q_z\mathrm{d}x^2 = 0\,.$$ (3.34)

If the bending moment on the right-hand face is expanded into a Taylor's series of first order similar to Eq. (3.30) and consideration that the term $\frac{1}{2}q_z\mathrm{d}x^2$ as infinitesimal small size of higher order can be disregarded, finally the following results:

$$\frac{\mathrm{d}M_y(x)}{\mathrm{d}x} = Q_z(x)\,.$$ (3.35)

The combination of Eqs. (3.32) and (3.35) leads to the relation between the bending moment and the distributed load:

$$\frac{\mathrm{d}^2 M_y(x)}{\mathrm{d}x^2} = \frac{\mathrm{d}Q_z(x)}{\mathrm{d}x} = -q_z(x)\,.$$ (3.36)

Finally, the elementary basic equations for the bending of a beam in the x-z plane for arbitrary moment loading $M_y(x)$ are summarized in Table 3.4.

Table 3.4 Elementary basic equations for the bending of a Bernoulli beam in the x-z plane. The differential equations are given under the assumption of constant bending stiffness EI_y

Name	Equation
Kinematics	$\varepsilon_x(x, z) = -z \dfrac{\mathrm{d}^2 u_z(x)}{\mathrm{d}x^2}$
Equilibrium	$\dfrac{\mathrm{d}Q_z(x)}{\mathrm{d}x} = -q_z(x); \quad \dfrac{\mathrm{d}M_y(x)}{\mathrm{d}x} = Q_z(x)$
Constitutive equation	$\sigma_x(x, z) = E\varepsilon_x(x, z)$
Stress	$\sigma_x(x, z) = \dfrac{M_y(x)}{I_y} z(x)$
Diff'equation	$EI_y \dfrac{\mathrm{d}^2 u_z(x)}{\mathrm{d}x^2} = -M_y(x)$
	$EI_y \dfrac{\mathrm{d}^3 u_z(x)}{\mathrm{d}x^3} = -Q_z(x)$
	$EI_y \dfrac{\mathrm{d}^4 u_z(x)}{\mathrm{d}x^4} = q_z(x)$

3.5 Differential Equation

Two-time differentiation of Eq. (3.25) and consideration of the relation between bending moment and distributed load according to Eq. (3.36) lead to the classical type of differential equation of the bending line,

$$\frac{\mathrm{d}^2}{\mathrm{d}x^2}\left(EI_y \frac{\mathrm{d}^2 u_z}{\mathrm{d}x^2}\right) = q_z , \tag{3.37}$$

which is also referred to as the bending line-distributed load relation. For a beam with constant bending stiffness EI_y along the beam axis, the following results:

$$EI_y \frac{\mathrm{d}^4 u_z}{\mathrm{d}x^4} = q_z . \tag{3.38}$$

The differential equation of the bending line can of course also be expressed through the bending moment or the shear force as

$$EI_y \frac{\mathrm{d}^2 u_z}{\mathrm{d}x^2} = -M_y \quad \text{or} \tag{3.39}$$

$$EI_y \frac{\mathrm{d}^3 u_z}{\mathrm{d}x^3} = -Q_z . \tag{3.40}$$

Table 3.5 Different formulations of the partial differential equation for a Bernoulli beam in the x-z plane (x-axis: right facing; z-axis: upward facing)

Configuration	Partial differential equation
E, I_y	$EI_y \dfrac{\mathrm{d}^4 u_z}{\mathrm{d}x^4} = 0$
$E(x), I_y(x)$	$\dfrac{\mathrm{d}^2}{\mathrm{d}x^2}\left(E(x) I_y(x) \dfrac{\mathrm{d}^2 u_z}{\mathrm{d}x^2} \right) = 0$
$q_z(x)$	$EI_y \dfrac{\mathrm{d}^4 u_z}{\mathrm{d}x^4} = q_z(x)$
$m_y(x)$	$EI_y \dfrac{\mathrm{d}^4 u_z}{\mathrm{d}x^4} = \dfrac{\mathrm{d}m_y(x)}{\mathrm{d}x}$
$k(x)$	$EI_y \dfrac{\mathrm{d}^4 u_z}{\mathrm{d}x^4} = -k(x) u_z$

Equations (3.39) and (3.40) can be also written in the more general form for variable bending stiffness:

$$E(x) I_y(x) \frac{\mathrm{d}^2 u_z}{\mathrm{d}x^2} = -M_y(x)\,, \tag{3.41}$$

$$\frac{\mathrm{d}u_z}{\mathrm{d}x}\left(E(x) I_y(x) \frac{\mathrm{d}^2 u_z}{\mathrm{d}x^2} \right) = -Q_z(x)\,. \tag{3.42}$$

Depending on the problem and the fact which distribution ($q_z(x)$, $M_y(x)$ or $Q_z(x)$) is easier to state, one may start from one of the three formulations to derive the displacement field $u_z(x)$.

Different formulations of the fourth order differential equation are collected in Table 3.5 where different types of loadings, geometry and bedding are differentiated. The last case in Table 3.5 refers to the elastic foundation of a beam which is also know in the literature as Winkler foundation [12]. The elastic foundation or Winkler foundation modulus k has in the case of beams[11] the unit of force per unit area.

If we replace the common formulation of the second order derivative, i.e. $\frac{\mathrm{d}^2 \ldots}{\mathrm{d}x^2}$, by a formal operator symbol, i.e. $\mathcal{L}_2(\ldots)$, the basic equations can be stated in a more formal way as given in Table 3.6.

---

[11] In the general case, the unit of the elastic foundation modulus is force per unit area per unit length, i.e. $\frac{\mathrm{N}}{\mathrm{m}^2}/\mathrm{m} = \frac{\mathrm{N}}{\mathrm{m}^3}$.

Table 3.6 Different formulations of the basic equations for an Euler–Bernoulli beam (bending in the x-z plane; x-axis along the principal beam axis). E: Young's modulus; I_y: second moment of area; q_x: length-specific distributed force; $\mathcal{L}_2 = \frac{d^2(...)}{dx^2}$: second-order derivative

Specific formulation	General formulation
Kinematics	
$\varepsilon_x(x, z) = -z\dfrac{d^2 u_z(x)}{dx^2}$	$\varepsilon_x(x, z) = -z\mathcal{L}_2\left(u_z(x)\right)$
$\kappa = -\dfrac{d^2 u_z(x)}{dx^2}$	$\kappa = -\mathcal{L}_2\left(u_z(x)\right)$
Constitution	
$\sigma_x(x, z) = E\varepsilon_x(x, z)$	$\sigma_x(x, z) = C\varepsilon_x(x, z)$
$M_y(x) = EI_y\kappa(x)$	$M_y(x) = D\kappa(x)$
Equilibrium	
$\dfrac{d^2 M_y(x)}{dx^2} + q_z(x) = 0$	$\mathcal{L}_2^{\mathrm{T}}\left(M_y(x)\right) + q_z(x) = 0$
PDE	
$\dfrac{d^2}{dx^2}\left(EI_y\dfrac{d^2 u_z(x)}{dx^2}\right) - q_z(x) = 0$	$\mathcal{L}_2^{\mathrm{T}}\left(D\mathcal{L}_2\left(u_z(x)\right)\right) - q_z(x) = 0$

References

1. Altenbach H, Altenbach J, Naumenko K (1998) Ebene Flächentragwerke: Grundlagen der Modellierung und Berechnung von Scheiben und Platten. Springer, Berlin
2. Boresi AP, Schmidt RJ (2003) Advanced mechanics of materials. Wiley, New York
3. Budynas RG (1999) Advanced strength and applied stress analysis. McGraw-Hill Book, Singapore
4. Gould PL (1988) Analysis of shells and plates. Springer, New York
5. Gross D, Hauger W, Schröder J, Wall WA (2009) Technische Mechanik 2: Elastostatik. Springer, Berlin
6. Hartmann F, Katz C (2007) Structural analysis with finite elements. Springer, Berlin
7. Heyman J (1998) Structural analysis: a historical approach. Cambridge University Press, Cambridge
8. Hibbeler RC (2008) Mechanics of materials. Prentice Hall, Singapore
9. Öchsner A (2014) Elasto-plasticity of frame structure elements: modeling and simulation of rods and beams. Springer, Berlin
10. Szabó I (2003) Einführung in die Technische Mechanik: Nach Vorlesungen István Szabó. Springer, Berlin
11. Timoshenko S, Woinowsky-Krieger S (1959) Theory of plates and shells. McGraw-Hill Book Company, New York
12. Winkler E (1867) Die Lehre von der Elasticität und Festigkeit mit besonderer Rücksicht auf ihre Anwendung in der Technik. H. Dominicus, Prag

Chapter 4
Timoshenko Beams

Abstract This chapter covers the continuum mechanical description of beam members under the additional influence of shear stresses. Based on the three basic equations of continuum mechanics, i.e., the kinematics relationship, the constitutive law and the equilibrium equation, the partial differential equations, which describe the physical problem, are derived.

4.1 Introduction

The general difference regarding the deformation of a beam with and without shear influence has already been discussed in Sect. 3.1. In this section, the shear influence on the deformation is considered with the help of the Timoshenko beam theory [12, 13]. Within the framework of the following remarks, the definition of the shear strain and the relation between shear force and shear stress will first be covered.

For the derivation of the equation for the shear strain in the x-z plane, the infinitesimal rectangular beam element $ABCD$, shown in Fig. 4.1, is considered, which deforms under the influence of a pure shear stress [7]. Here, a change of the angle of the original right angles as well as a change in the lengths of the edges occurs.

The deformation of the point A can be described via the displacement fields $u_x(x, z)$ and $u_z(x, z)$. These two functions of *two* variables can be expanded in Taylor's series[1] of first order around point A to approximately calculate the deformations of the points B and D:

---

[1] For a function $f(x, z)$ of two variables usually a Taylor's series expansion of first order is formulated around the point (x_0, z_0) as follows: $f(x, z) = f(x_0 + \mathrm{d}x, z_0 + \mathrm{d}z) \approx f(x_0, z_0) + \left(\frac{\partial f}{\partial x}\right)_{x_0, z_0} \times (x - x_0) + \left(\frac{\partial f}{\partial z}\right)_{x_0, z_0} \times (z - z_0)$.

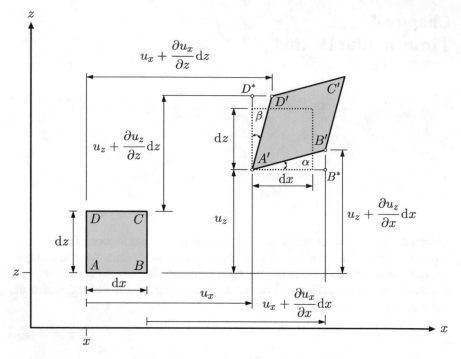

Fig. 4.1 Definition of the shear strain γ_{xz} in the x-z plane at an infinitesimal beam element

$$u_{x,B} = u_x(x + \mathrm{d}x, z) = u_x(x, z) + \frac{\partial u_x}{\partial x}\mathrm{d}x + \frac{\partial u_x}{\partial z}\mathrm{d}z \,, \tag{4.1}$$

$$u_{z,B} = u_z(x + \mathrm{d}x, z) = u_z(x, z) + \frac{\partial u_z}{\partial x}\mathrm{d}x + \frac{\partial u_z}{\partial z}\mathrm{d}z \,, \tag{4.2}$$

or alternatively

$$u_{x,D} = u_x(x, z + \mathrm{d}z) = u_x(x, z) + \frac{\partial u_x}{\partial x}\mathrm{d}x + \frac{\partial u_x}{\partial z}\mathrm{d}z \,, \tag{4.3}$$

$$u_{z,D} = u_z(x, z + \mathrm{d}z) = u_z(x, z) + \frac{\partial u_z}{\partial x}\mathrm{d}x + \frac{\partial u_z}{\partial z}\mathrm{d}z \,. \tag{4.4}$$

In Eqs. (4.1) up to (4.4), $u_x(x, z)$ and $u_z(x, z)$ represent the so-called rigid-body displacements, which do not cause a deformation. If one considers that point B has the coordinates $(x + \mathrm{d}x, z)$ and D the coordinates $(x, z + \mathrm{d}z)$, the following results:

$$u_{x,B} = u_x(x, z) + \frac{\partial u_x}{\partial x}\mathrm{d}x \,, \tag{4.5}$$

$$u_{z,B} = u_z(x, z) + \frac{\partial u_z}{\partial x}\mathrm{d}x \,, \tag{4.6}$$

or alternatively

$$u_{x,D} = u_x(x, z) + \frac{\partial u_x}{\partial z} dz, \qquad (4.7)$$

$$u_{z,D} = u_z(x, z) + \frac{\partial u_z}{\partial z} dz. \qquad (4.8)$$

The total shear strain γ_{xz} of the deformed beam element $A'B'C'D'$ results, according to Fig. 4.1, from the sum of the angles α and β. The two angles can be identified in the rectangle, which is deformed to a rhombus. Under consideration of the two right-angled triangles $A'D^*D'$ and $A'B^*B'$, these two angles can be expressed as:

$$\tan \alpha = \frac{\dfrac{\partial u_z}{\partial x} dx}{dx + \dfrac{\partial u_x}{\partial x} dx} \quad \text{and} \quad \tan \beta = \frac{\dfrac{\partial u_x}{\partial z} dz}{dz + \dfrac{\partial u_z}{\partial z} dz}. \qquad (4.9)$$

It holds approximately for small deformations that $\tan \alpha \approx \alpha$ and $\tan \beta \approx \beta$ or alternatively $\frac{\partial u_x}{\partial x} \ll 1$ and $\frac{\partial u_z}{\partial z} \ll 1$, so that the following expression results for the shear strain:

$$\gamma_{xz} = \alpha + \beta \approx \frac{\partial u_z}{\partial x} + \frac{\partial u_x}{\partial z}. \qquad (4.10)$$

This total change of the angle is also called the engineering shear strain. In contrast to this, the expression $\varepsilon_{xz} = \frac{1}{2} \gamma_{xz} = \frac{1}{2}(\frac{\partial u_z}{\partial x} + \frac{\partial u_x}{\partial z})$ is known as the tensorial definition (tensor shear strain) in the literature [16]. Due to the symmetry of the strain tensor, the identity $\gamma_{ij} = \gamma_{ji}$ applies to the tensor elements outside the main diagonal.

The algebraic sign of the shear strain needs to be explained in the following with the help of Fig. 4.2 for the special case that only one shear force acts in parallel to the z-axis. If a shear force acts in the direction of the positive z-axis at the right-hand face—hence a positive shear force distribution is being assumed at this point—, according to Fig. 4.2a under consideration of Eq. (4.10) a positive shear strain results.

Fig. 4.2 Definition of a **a** positive and **b** negative shear strain in the x-z plane

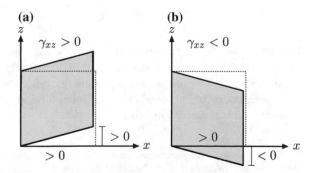

Fig. 4.3 Shear stress distribution: **a** real distribution for a rectangular cross section and **b** Timoshenko's approximation

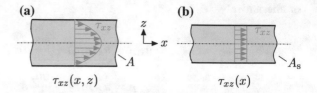

In a similar way, a negative shear force distribution leads to a negative shear strain according to Fig. 4.2b.

It has already been mentioned in Sect. 3.1 that the shear stress distribution is variable over the cross-section. As an example, the parabolic shear stress distribution was illustrated over a rectangular cross section in Fig. 3.3. Based on Hooke's law for a one-dimensional shear stress state, it can be derived that the shear strain has to exhibit a corresponding parabolic course. From the shear stress distribution in the cross-sectional area at location x of the beam,[2] one receives the acting shear force through integration as:

$$Q_z = \int_A \tau_{xz}(y, z)\, \mathrm{d}A\,. \tag{4.11}$$

However, to simplify the problem, it is assumed for the Timoshenko beam that an equivalent *constant* shear stress and strain act, see Fig. 4.3:

$$\tau_{xz}(y, z) \rightarrow \tau_{xz}\,. \tag{4.12}$$

This constant shear stress results from the shear force, which acts in an equivalent cross-sectional area, the so-called shear area A_s:

$$\tau_{xz} = \frac{Q_z}{A_s}\,, \tag{4.13}$$

whereupon the relation between the shear area A_s and the actual cross-sectional area A is referred to as the shear correction factor k_s:

$$k_s = \frac{A_s}{A}\,. \tag{4.14}$$

Different assumptions can be made to calculate the shear correction factor [3]. As an example, it can be demanded [1] that the elastic strain energy of the equivalent shear stress has to be identical with the energy, which results from the acting shear stress distribution in the actual cross-sectional-area. A comparison for a rectangular cross section is presented in Table 4.1.

---

[2]A closer analysis of the shear stress distribution in the cross-sectional area shows that the shear stress does not just alter over the height of the beam but also through the width of the beam. If the width of the beam is small when compared to the height, only a small change along the width occurs and one can assume in the first approximation a constant shear stress throughout the width: $\tau_{xz}(y, z) \rightarrow \tau_{xz}(z)$. See for example [2, 15].

Table 4.1 Comparison of shear correction factor values for a rectangular cross section based on different approaches

k_s	Comment	Reference
$\frac{2}{3}$	–	[12, 14]
$0.833\left(=\frac{5}{6}\right)$	$\nu = 0.0$	[3]
0.850	$\nu = 0.3$	
0.870	$\nu = 0.5$	

Table 4.2 Characteristics of different cross sections in the y-z plane. I_y and I_z: axial second moment of area; A: cross-sectional area; k_s: shear correction factor. Adapted from [17]

Cross-section	I_y	I_z	A	k_s
	$\dfrac{\pi R^4}{4}$	$\dfrac{\pi R^4}{4}$	πR^2	$\dfrac{9}{10}$
	$\pi R^3 t$	$\pi R^3 t$	$2\pi R t$	0.5
	$\dfrac{bh^3}{12}$	$\dfrac{hb^3}{12}$	hb	$\dfrac{5}{6}$
	$\dfrac{h^2}{6}(ht_w + 3bt_f)$	$\dfrac{b^2}{6}(bt_f + 3ht_w)$	$2(bt_f + ht_w)$	$\dfrac{2ht_w}{A}$
	$\dfrac{h^2}{12}(ht_w + 6bt_f)$	$\dfrac{b^3 t_f}{6}$	$ht_w + 2bt_f$	$\dfrac{ht_w}{A}$

Different geometric characteristics of simple geometric cross-sections—including the shear correction factor[3]—are collected in Table 4.2 [4, 17]. Further details regarding the shear correction factor for arbitrary cross-sections can be taken from [5].

It is obvious that the equivalent constant shear stress can alter along the center line of the beam, in case the shear force along the center line of the beam changes. The attribute 'constant' thus just refers to the cross-sectional area at location x and the equivalent constant shear stress is therefore in general a function of the coordinate of length for the Timoshenko beam:

$$\tau_{xz} = \tau_{xz}(x) \,. \tag{4.15}$$

The so-called Timoshenko beam can be generated by superposing a shear deformation on a Bernoulli beam according to Fig. 4.4.

One can see that the Bernoulli hypothesis is partly no longer fulfilled for the Timoshenko beam: Plane cross sections remain plane after the deformation. However, a cross section which stood at right angles on the beam axis before the deformation is not at right angles on the beam axis after the deformation. If the demand for planeness of the cross sections is also given up, one reaches theories of higher-order [6, 9, 10], at which, for example, a parabolic course of the shear strain and stress in the displacement field are considered, see Fig. 4.5. Therefore, a shear correction factor is not required for these theories of higher-order.

4.2   Kinematics

According to the alternative derivation in Sect. 3.2, the kinematics relation can also be derived for the beam with shear action, by considering the angle ϕ_y instead of the angle φ_y, see Figs. 4.4c and 4.6.

Following an equivalent procedure as in Sect. 3.2, the corresponding relationships are obtained:

$$\sin \phi_y = \frac{u_x}{z} \approx \phi_y \ \text{ or } \ u_x = +z\phi_y \,, \tag{4.16}$$

wherefrom, via the general relation for the strain, meaning $\varepsilon_x = \mathrm{d}u_x/\mathrm{d}x$, the kinematics relation results through differentiation with respect to the x-coordinate:

$$\varepsilon_x = +z\frac{\mathrm{d}\phi_y}{\mathrm{d}x} \,. \tag{4.17}$$

Note that $\phi_y \to \varphi_y = -\frac{\mathrm{d}u_z}{\mathrm{d}x}$ results from neglecting the shear deformation and a relation according to Eq. (3.16) results as a special case. Furthermore, the following

---

[3]It should be noted that the so-called form factor for shear is also known in the literature. This results as the reciprocal of the shear correction factor.

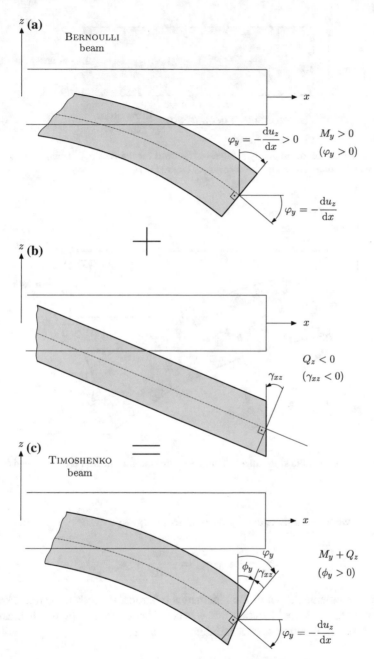

Fig. 4.4 Superposition of the Bernoulli beam (**a**) and the shear deformation (**b**) to the Timoshenko beam (**c**) in the x-z plane. Note that the deformation is exaggerated for better illustration

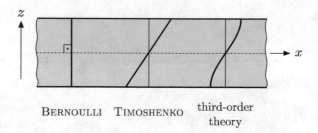

BERNOULLI   TIMOSHENKO   third-order
theory

Fig. 4.5 Deformation of originally plane cross sections for the Bernoulli beam (left), the Timoshenko beam (middle) and a higher-order theory (right) [11]

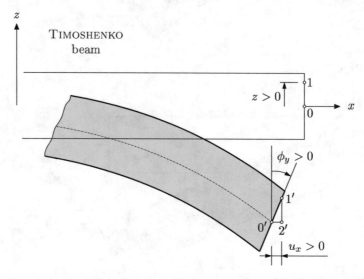

Fig. 4.6 Derivation of the kinematics relation. Note that the deformation is exaggerated for better illustration

relation between the angles can be derived from Fig. 4.4c

$$\phi_y = \varphi_y + \gamma_{xz} = -\frac{\mathrm{d}u_z}{\mathrm{d}x} + \gamma_{xz}, \tag{4.18}$$

which complements the set of the kinematics relations. It needs to be remarked that at this point the so-called bending line was considered. Therefore, the displacement field u_z is only a function of *one* variable: $u_z = u_z(x)$.

4.3 Equilibrium

The derivation of the equilibrium condition for the Timoshenko beam is identical with the derivation for the Bernoulli beam according to Sect. 3.4:

$$\frac{dQ_z(x)}{dx} = -q_z(x), \tag{4.19}$$

$$\frac{dM_y(x)}{dx} = +Q_z(x). \tag{4.20}$$

4.4 Constitution

For the consideration of the constitutive relation, Hooke's law for a one-dimensional normal stress state and for a one-dimensional shear stress state is used [8]:

$$\sigma_x = E\varepsilon_x, \tag{4.21}$$

$$\tau_{xz} = G\gamma_{xz}, \tag{4.22}$$

whereupon the shear modulus G can be calculated based on the Young's modulus E and the Poisson's ratio ν as:

$$G = \frac{E}{2(1+\nu)}. \tag{4.23}$$

According to the equilibrium configuration of Fig. 3.9 and Eq. (3.22), the relation between the internal moment and the bending stress can be used for the Timoshenko beam as follows:

$$dM_y = (+z)(+\sigma_x)dA, \tag{4.24}$$

or alternatively after integration under the consideration of the constitutive equation (4.21) and the kinematics relation (4.17):

$$M_y(x) = +EI_y\frac{d\phi_y(x)}{dx}. \tag{4.25}$$

The relation between shear force and cross-sectional rotation results from the equilibrium equation (4.20) as:

$$Q_z(x) = +\frac{dM_y(x)}{dx} = +EI_y\frac{d^2\phi_y(x)}{dx^2}. \tag{4.26}$$

Before looking in more detail at the differential equations of the bending line, let us summarize the basic equations for the Timoshenko beam in Table 4.3. Note that the

Table 4.3 Elementary basic equations for the bending of a Timoshenko beam in the x-z plane (x-axis: right facing; z-axis: upward facing)

Relation	Equation
Kinematics	$\varepsilon_x(x, z) = +z\dfrac{\mathrm{d}\phi_y(x)}{\mathrm{d}x}$ and $\phi_y(x) = -\dfrac{\mathrm{d}u_z(x)}{\mathrm{d}x} + \gamma_{xz}(x)$
Equilibrium	$\dfrac{\mathrm{d}Q_z(x)}{\mathrm{d}x} = -q_z(x)\,;\ \dfrac{\mathrm{d}M_y(x)}{\mathrm{d}x} = +Q_z(x)$
Constitution	$\sigma_x(x, z) = E\varepsilon_x(x, z)$ and $\tau_{xz}(x) = G\gamma_{xz}(x)$

normal stress and normal strain are functions of both spatial coordinates, i.e. x and z. However, the shear stress and shear strain are only dependent on the x-coordinate, since an equivalent *constant* shear stress has been introduced over the cross section as an approximation of the Timoshenko beam theory.

4.5 Differential Equation

Within the previous section, the relation between the internal moment and the cross-sectional rotation was derived from the normal stress distribution with the help of Hooke's law, see Eq. (4.25). Differentiation of this relation with respect to the x-coordinate leads to the following expression

$$\frac{\mathrm{d}M_y}{\mathrm{d}x} = \frac{\mathrm{d}}{\mathrm{d}x}\left(EI_y\frac{\mathrm{d}\phi_y}{\mathrm{d}x}\right), \tag{4.27}$$

which can be transformed with the help of the equilibrium relation (4.20), the constitutive equation (4.22), and the relation for the shear stress according to (4.13) and (4.14) to

$$\frac{\mathrm{d}}{\mathrm{d}x}\left(EI_y\frac{\mathrm{d}\phi_y}{\mathrm{d}x}\right) = +k_s GA\gamma_{xz}. \tag{4.28}$$

If the kinematics relation (4.18) is considered in the last equation, the so-called bending differential equation results in:

$$\frac{\mathrm{d}}{\mathrm{d}x}\left(EI_y\frac{\mathrm{d}\phi_y}{\mathrm{d}x}\right) - k_s GA\left(\frac{\mathrm{d}u_z}{\mathrm{d}x} + \phi_y\right) = 0. \tag{4.29}$$

Considering the shear stress according to Eqs. (4.13) and (4.14) in the expression of Hooke's law according to (4.22), one obtains

$$Q_z = k_s AG\gamma_{xz}. \tag{4.30}$$

Introducing the equilibrium relation (4.20) and the kinematics relation (4.18) in the last equation gives:

$$\frac{dM_y}{dx} = +k_s AG\left(\frac{du_z}{dx} + \phi_y\right). \tag{4.31}$$

After differentiation and the consideration of the equilibrium relations according to Eqs. (4.19) and (4.20), the so-called shear differential equation results finally in:

$$\frac{d}{dx}\left[k_s AG\left(\frac{du_z}{dx} + \phi_y\right)\right] = -q_z(x). \tag{4.32}$$

Therefore, the shear flexible Timoshenko beam is described through the following two coupled differential equations of second order:

$$\frac{d}{dx}\left(EI_y\frac{d\phi_y}{dx}\right) - k_s AG\left(\frac{du_z}{dx} + \phi_y\right) = 0, \tag{4.33}$$

$$\frac{d}{dx}\left[k_s AG\left(\frac{du_z}{dx} + \phi_y\right)\right] = -q_z(x). \tag{4.34}$$

This system contains two unknown functions, namely the deflection $u_z(x)$ and the cross-sectional rotation $\phi_y(x)$. Boundary conditions must be formulated for both functions to be able to solve the system of differential equations for a specific problem.

Different formulations of these coupled differential equations are collected in Table 4.4 where different types of loadings, geometry and bedding are differentiated. The last case in Table 4.4 refers again to the elastic or Winkler foundation of a beam, [18]. The elastic foundation modulus k has in the case of beams the unit of force per unit area.

A single-equation description for the Timoshenko beam can be obtained under the assumption of constant material (E, G) and geometrical (I_z, A, k_s) properties: Rearranging and two-times differentiation of Eq. (4.34) gives:

$$\frac{d\phi_y}{dx} = -\frac{d^2 u_z}{dx^2} - \frac{q_z}{k_s GA}, \tag{4.35}$$

$$\frac{d^3 \phi_y}{dx^3} = -\frac{d^4 u_z}{dx^4} - \frac{d^2 q_z}{k_s GA dx^2}. \tag{4.36}$$

One-time differentiation of Eq. (4.33) gives:

$$EI_y\frac{d^3 \phi_y}{dx^3} - k_s AG\left(\frac{d^2 u_z}{dx^2} + \frac{d\phi_y}{dx}\right) = 0. \tag{4.37}$$

Table 4.4 Different formulations of the partial differential equation for a Timoshenko beam in the x-z plane (x-axis: right facing; z-axis: upward facing)

Configuration	Partial differential equation
$E, I_y, A, G, k_\mathrm{s}$	$EI_y \dfrac{\mathrm{d}^2\phi_y}{\mathrm{d}x^2} - k_\mathrm{s}GA\left(\dfrac{\mathrm{d}u_z}{\mathrm{d}x} + \phi_y\right) = 0$ $\;$ $k_\mathrm{s}GA\left(\dfrac{\mathrm{d}^2u_z}{\mathrm{d}x^2} + \dfrac{\mathrm{d}\phi_y}{\mathrm{d}x}\right) = 0$
$E(x), I_y(x)$ $k_\mathrm{s}(x), A(x), G(x)$	$\dfrac{\mathrm{d}}{\mathrm{d}x}\left(E(x)I_y(x)\dfrac{\mathrm{d}\phi_y}{\mathrm{d}x}\right) - k_\mathrm{s}(x)G(x)A(x)\left(\dfrac{\mathrm{d}u_z}{\mathrm{d}x} + \phi_y\right) = 0$ $\;$ $\dfrac{\mathrm{d}}{\mathrm{d}x}\left[k_\mathrm{s}(x)G(x)A(x)\left(\dfrac{\mathrm{d}u_z}{\mathrm{d}x} + \phi_y\right)\right] = 0$
$q_z(x)$	$EI_y \dfrac{\mathrm{d}^2\phi_y}{\mathrm{d}x^2} - k_\mathrm{s}GA\left(\dfrac{\mathrm{d}u_z}{\mathrm{d}x} + \phi_y\right) = 0$ $\;$ $k_\mathrm{s}GA\left(\dfrac{\mathrm{d}^2u_z}{\mathrm{d}x^2} + \dfrac{\mathrm{d}\phi_y}{\mathrm{d}x}\right) = -q_z(x)$
$m_y(x)$	$EI_y \dfrac{\mathrm{d}^2\phi_y}{\mathrm{d}x^2} - k_\mathrm{s}GA\left(\dfrac{\mathrm{d}u_z}{\mathrm{d}x} + \phi_y\right) = -m_y(x)$ $\;$ $k_\mathrm{s}GA\left(\dfrac{\mathrm{d}^2u_z}{\mathrm{d}x^2} + \dfrac{\mathrm{d}\phi_y}{\mathrm{d}x}\right) = 0$
$k(x)$	$EI_y \dfrac{\mathrm{d}^2\phi_y}{\mathrm{d}x^2} - k_\mathrm{s}GA\left(\dfrac{\mathrm{d}u_z}{\mathrm{d}x} + \phi_y\right) = 0$ $\;$ $k_\mathrm{s}GA\left(\dfrac{\mathrm{d}^2u_z}{\mathrm{d}x^2} + \dfrac{\mathrm{d}\phi_y}{\mathrm{d}x}\right) = k(x)u_z$

Inserting Eq. (4.35) into (4.37) and consideration of (4.36) gives finally the following expression:

$$EI_y \frac{\mathrm{d}^4u_z(x)}{\mathrm{d}x^4} = q_z(x) - \frac{EI_y}{k_\mathrm{s}AG}\frac{\mathrm{d}^2q_z(x)}{\mathrm{d}x^2}. \tag{4.38}$$

The last equation reduces for shear-rigid beams, i.e. $k_\mathrm{s}AG \to \infty$, to the classical Bernoulli formulation as given in Table 3.5.

For the derivation of analytical solutions, the system of coupled differential equations as summarized in Table 4.4 has to be solved.

Table 4.5 Different formulations of the basic equations for a Timoshenko beam (bending in the x-z plane; x-axis along the principal beam axis). E: Young's modulus; G: shear modulus; A: cross-sectional area; I_y: second moment of area; k_s: shear correction factor; q_z: length-specific distributed force; m_z: length-specific distributed moment; e: generalized strains; s: generalized stresses

Specific formulation	General formulation
Kinematics	
$\begin{bmatrix} \frac{du_z}{dx} + \phi_y \\ \frac{d\phi_y}{dx} \end{bmatrix} = \begin{bmatrix} \frac{d}{dx} & 1 \\ 0 & \frac{d}{dx} \end{bmatrix} \begin{bmatrix} u_z \\ \phi_y \end{bmatrix}$	$e = \mathcal{L}_1 u$
Constitution	
$\begin{bmatrix} -Q_z \\ M_y \end{bmatrix} = \begin{bmatrix} -k_s AG & 0 \\ 0 & EI_y \end{bmatrix} \begin{bmatrix} \frac{du_z}{dx} + \phi_y \\ \frac{d\phi_y}{dx} \end{bmatrix}$	$s = De$
Equilibrium	
$\begin{bmatrix} \frac{d}{dx} & 0 \\ 1 & \frac{d}{dx} \end{bmatrix} \begin{bmatrix} -Q_z \\ M_y \end{bmatrix} + \begin{bmatrix} -q_z \\ +m_z \end{bmatrix} = \begin{bmatrix} 0 \\ 0 \end{bmatrix}$	$\mathcal{L}_1^{\mathrm{T}} s + b = 0$
PDE	
$-\dfrac{d}{dx}\left[k_s GA\left(\dfrac{du_z}{dx} + \phi_y \right) \right] - q_z = 0$ $\dfrac{d}{dx}\left(EI_y \dfrac{d\phi_y}{dx} \right) - k_s GA\left(\dfrac{du_z}{dx} + \phi_y \right) + m_y = 0$,	$\mathcal{L}_1^{\mathrm{T}} D\mathcal{L}_1 u + b = 0$

If we replace in the previous formulations the first order derivative, i.e. $\frac{d...}{dx}$, by a formal operator symbol, i.e. the $\mathcal{L}_1$–matrix, then the basic equations of the Timoshenko beam can be stated in a more formal way as given in Table 4.5.

It should be noted here that the formulation of the partial differential equation in Table 4.5 is slightly different to Eqs. (4.33) and (4.34): First of all, we considered here a distributed moment m_z. Furthermore, the first PDE in Table 4.5 is multiplied by -1 compared to Eq. (4.34). This must be carefully considered as soon as finite element derivations based on both approaches are compared.

Similar to the end of Sect. 3.3, it is more advantageous in a general approach to work with the generalized stresses $s = \begin{bmatrix} -Q_z, & M_y \end{bmatrix}^{\mathrm{T}}$ and generalized strains $e = \begin{bmatrix} \frac{du_z}{dx} + \phi_y, & \frac{d\phi_y}{dx} \end{bmatrix}^{\mathrm{T}} = \begin{bmatrix} \gamma_{xz}, & \kappa_y \end{bmatrix}^{\mathrm{T}}$ since these quantities do not depend on the vertical coordinate z. Classical stress and strain values are changing along the vertical coordinate. The representation of the constitutive relationship based on the classical stress-strain quantities and the corresponding generalized quantities is represented in Fig. 4.7.

The general formulations of the basic equations for a Timoshenko beam as given in Table 4.5 can be slightly modified to avoid some esthetic appeals. The representations of generalized stresses, generalized stiffness matrix, and distributed loads contain a minus sign. Let us start with the constitutive equation, i.e.,

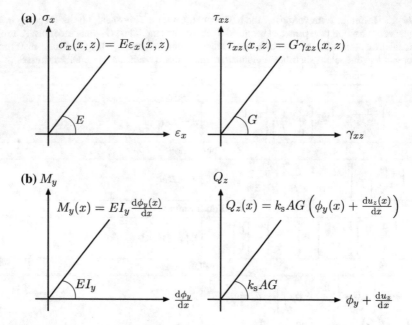

Fig. 4.7 Formulation of the constitutive law based on **a** classical stress-strain and **b** generalized-stress-generalized-strain relations

$$\begin{bmatrix} -Q_z \\ M_y \end{bmatrix} = \begin{bmatrix} -k_s AG & 0 \\ 0 & EI_y \end{bmatrix} \begin{bmatrix} \frac{du_z}{dx} + \phi_y \\ \frac{d\phi_y}{dx} \end{bmatrix}, \tag{4.39}$$

or under elimination of the minus sign in the mentioned matrices:

$$\begin{bmatrix} -1 & 0 \\ 0 & 1 \end{bmatrix} \begin{bmatrix} Q_z \\ M_y \end{bmatrix} = \begin{bmatrix} -1 & 0 \\ 0 & 1 \end{bmatrix} \begin{bmatrix} k_s AG & 0 \\ 0 & EI_y \end{bmatrix} \begin{bmatrix} \frac{du_z}{dx} + \phi_y \\ \frac{d\phi_y}{dx} \end{bmatrix}. \tag{4.40}$$

The diagonal matrix $\lceil -1 \ 1 \rfloor$ can be eliminated from the last equation to obtain the modified constitutive law in matrix form:

$$\begin{bmatrix} Q_z \\ M_y \end{bmatrix} = \begin{bmatrix} k_s AG & 0 \\ 0 & EI_y \end{bmatrix} \begin{bmatrix} \frac{du_z}{dx} + \phi_y \\ \frac{d\phi_y}{dx} \end{bmatrix}. \tag{4.41}$$

The next step is to have a closer look on the equilibrium equation, i.e.,

$$\begin{bmatrix} \frac{d}{dx} & 0 \\ 1 & \frac{d}{dx} \end{bmatrix} \begin{bmatrix} -Q_z \\ M_y \end{bmatrix} + \begin{bmatrix} -q_z \\ +m_z \end{bmatrix} = \begin{bmatrix} 0 \\ 0 \end{bmatrix}, \tag{4.42}$$

or again re-written based on the diagonal matrix:

Table 4.6 Alternative formulations of the basic equations for a Timoshenko beam (bending occurs in the x-z plane)

Specific formulation	General formulation
Kinematics	
$\begin{bmatrix} \frac{du_z}{dx} + \phi_y \\ \frac{d\phi_y}{dx} \end{bmatrix} = \begin{bmatrix} \frac{d}{dx} & 1 \\ 0 & \frac{d}{dx} \end{bmatrix} \begin{bmatrix} u_z \\ \phi_y \end{bmatrix}$	$e = \mathcal{L}_1 u$
Constitution	
$\begin{bmatrix} Q_z \\ M_y \end{bmatrix} = \begin{bmatrix} k_s AG & 0 \\ 0 & EI_y \end{bmatrix} \begin{bmatrix} \frac{du_z}{dx} + \phi_y \\ \frac{d\phi_y}{dx} \end{bmatrix}$	$s^* = D^* e$
Equilibrium	
$\begin{bmatrix} \frac{d}{dx} & 0 \\ -1 & \frac{d}{dx} \end{bmatrix} \begin{bmatrix} Q_z \\ M_y \end{bmatrix} + \begin{bmatrix} q_z \\ +m_z \end{bmatrix} = \begin{bmatrix} 0 \\ 0 \end{bmatrix}$	$\mathcal{L}_{1*}^T s^* + b^* = 0$
PDE	
$\frac{d}{dx}\left[k_s GA \left(\frac{du_z}{dx} + \phi_y \right) \right] + q_z = 0$ $\frac{d}{dx}\left(EI_y \frac{d\phi_y}{dx} \right) - k_s GA \left(\frac{du_z}{dx} + \phi_y \right) + m_y = 0,$	$\mathcal{L}_{1*}^T D^* \mathcal{L}_1 u + b^* = 0$

$$\begin{bmatrix} \frac{d}{dx} & 0 \\ 1 & \frac{d}{dx} \end{bmatrix} \begin{bmatrix} -1 & 0 \\ 0 & 1 \end{bmatrix} \begin{bmatrix} Q_z \\ M_y \end{bmatrix} + \begin{bmatrix} -1 & 0 \\ 0 & 1 \end{bmatrix} \begin{bmatrix} q_z \\ m_z \end{bmatrix} = \begin{bmatrix} 0 \\ 0 \end{bmatrix}, \qquad (4.43)$$

$$\begin{bmatrix} -\frac{d}{dx} & 0 \\ -1 & \frac{d}{dx} \end{bmatrix} \begin{bmatrix} Q_z \\ M_y \end{bmatrix} + \begin{bmatrix} -1 & 0 \\ 0 & 1 \end{bmatrix} \begin{bmatrix} q_z \\ m_z \end{bmatrix} = \begin{bmatrix} 0 \\ 0 \end{bmatrix}. \qquad (4.44)$$

Let us now multiply the last equation with the diagonal matrix from the left-hand side:

$$\begin{bmatrix} -1 & 0 \\ 0 & 1 \end{bmatrix} \begin{bmatrix} -\frac{d}{dx} & 0 \\ -1 & \frac{d}{dx} \end{bmatrix} \begin{bmatrix} Q_z \\ M_y \end{bmatrix} + \underbrace{\begin{bmatrix} -1 & 0 \\ 0 & 1 \end{bmatrix} \begin{bmatrix} -1 & 0 \\ 0 & 1 \end{bmatrix}}_{I} \begin{bmatrix} q_z \\ m_z \end{bmatrix} = \begin{bmatrix} -1 & 0 \\ 0 & 1 \end{bmatrix} \begin{bmatrix} 0 \\ 0 \end{bmatrix}, \qquad (4.45)$$

Or finally as the modified expression of the equilibrium equation:

$$\begin{bmatrix} \frac{d}{dx} & 0 \\ -1 & \frac{d}{dx} \end{bmatrix} \begin{bmatrix} Q_z \\ M_y \end{bmatrix} + \begin{bmatrix} q_z \\ m_z \end{bmatrix} = \begin{bmatrix} 0 \\ 0 \end{bmatrix}. \qquad (4.46)$$

The modified basic equations, i.e., 'without the minus sign', are summarized in Table 4.6.

References

1. Bathe K-J (1996) Finite element procedures. Prentice-Hall, Upper Saddle River
2. Beer FP, Johnston ER Jr, DeWolf JT, Mazurek DF (2009) Mechanics of materials. McGraw-Hill, New York
3. Cowper GR (1966) The shear coefficient in Timoshenko's beam theory. J Appl Mech 33:335–340
4. Gere JM, Timoshenko SP (1991) Mechanics of materials. PWS-KENT Publishing Company, Boston
5. Gruttmann F, Wagner W (2001) Shear correction factors in Timoshenko's beam theory for arbitrary shaped cross-sections. Comput Mech 27:199–207
6. Levinson M (1981) A new rectangular beam theory. J Sound Vib 74:81–87
7. Öchsner A (2014) Elasto-plasticity of frame structure elements: modeling and simulation of rods and beams. Springer, Berlin
8. Öchsner A (2016) Continuum damage and fracture mechanics. Springer, Singapore
9. Reddy JN (1984) A simple higher-order theory for laminated composite plate. J Appl Mech 51:745–752
10. Reddy JN (1997) Mechanics of laminated composite plates: theory and analysis. CRC Press, Boca Raton
11. Reddy JN (1997) On locking-free shear deformable beam finite elements. Comput Method Appl M 149:113–132
12. Timoshenko SP (1921) On the correction for shear of the differential equation for transverse vibrations of prismatic bars. Philos Mag 41:744–746
13. Timoshenko SP (1922) On the transverse vibrations of bars of uniform cross-section. Philos Mag 43:125–131
14. Timoshenko S (1940) Strength of materials - part I elementary theory and problems. D. Van Nostrand Company, New York
15. Timoshenko SP, Goodier JN (1970) Theory of elasticity. McGraw-Hill, New York
16. Twiss RJ, Moores EM (1992) Structural geology. WH Freeman & Co, New York
17. Weaver W Jr, Gere JM (1980) Matrix analysis of framed structures. Van Nostrand Reinhold Company, New York
18. Winkler E (1867) Die Lehre von der Elasticität und Festigkeit mit besonderer Rücksicht auf ihre Anwendung in der Technik. H. Dominicus, Prag

Chapter 5
Plane Members

Abstract This chapter covers the continuum mechanical description of plane elasticity members. Based on the three basic equations of continuum mechanics, i.e., the kinematics relationship, the constitutive law and the equilibrium equation, the partial differential equation, which describes the physical problem, is derived.

5.1 Introduction

A plane elasticity element is defined as a thin two-dimensional member, as schematically shown in Fig. 5.1, with a much smaller thickness t than the planar dimensions. It can be seen as a two-dimensional extension or generalization of the rod. The following derivations are restricted to some simplifications:

- the thickness t is constant and much smaller than the planer dimensions a and b,
- the undeformed member shape is planar,
- the material is isotropic, homogenous and linear-elastic according to Hooke's law for a plane stress or plane strain state,
- external forces act only at the boundary parallel to the plane of the member,
- external forces are distributed uniformly over the thickness, and
- only rectangular members are considered.

The analogies between the rod and plane elasticity theories are summarized in Table 5.1.

5.2 Kinematics

The kinematics or strain-displacement relations extract the strain field contained in a displacement field. Using engineering definitions of strain, the following relations can be obtained [1, 2]:

© The Author(s), under exclusive license to Springer Nature Switzerland AG 2020   47
A. Öchsner, *Partial Differential Equations of Classical Structural Members*,
SpringerBriefs in Continuum Mechanics,
https://doi.org/10.1007/978-3-030-35311-7_5

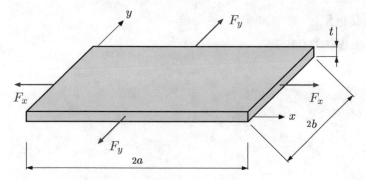

Fig. 5.1 General configuration for a plane elasticity problem

Table 5.1 Difference between rod, beam and plane element

Rod	Beam	Plane element
1D	1D	2D
Deformation along principal axis	Deformation perpendicular to principal axis	In-plane deformation
u_x	u_z, φ_y	u_x, u_y

$$\varepsilon_x = \frac{\partial u_x}{\partial x} ; \ \varepsilon_y = \frac{\partial u_y}{\partial y} ; \ \gamma_{xy} = 2\varepsilon_{xy} = \frac{\partial u_x}{\partial y} + \frac{\partial u_y}{\partial x} . \tag{5.1}$$

In matrix notation, these three relationships can be written as

$$\begin{bmatrix} \varepsilon_x \\ \varepsilon_y \\ 2\varepsilon_{xy} \end{bmatrix} = \begin{bmatrix} \frac{\partial}{\partial x} & 0 \\ 0 & \frac{\partial}{\partial y} \\ \frac{\partial}{\partial y} & \frac{\partial}{\partial x} \end{bmatrix} \begin{bmatrix} u_x \\ u_y \end{bmatrix} , \tag{5.2}$$

or symbolically as

$$\varepsilon = \mathcal{L}_1 u , \tag{5.3}$$

where $\mathcal{L}_1$ is the differential operator matrix.

5.3   Constitution

5.3.1   Plane Stress Case

The two-dimensional plane stress case ($\sigma_z = \sigma_{yz} = \sigma_{xz} = 0$) shown in Fig. 5.2 is commonly used for the analysis of thin, flat plates loaded in the plane of the plate (x-y plane) [3].

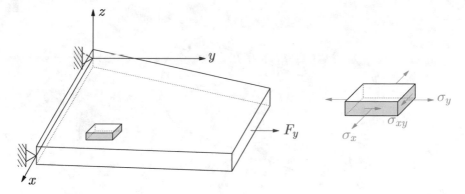

Fig. 5.2 Two-dimensional problem: plane stress case

It should be noted here that the normal thickness stress is zero ($\sigma_z = 0$) whereas the thickness normal strain is present ($\varepsilon_z \neq 0$).

The plane stress Hooke's law for a linear-elastic isotropic material based on the Young's modulus E and Poisson's ratio ν can be written for a constant temperature as

$$\begin{bmatrix} \sigma_x \\ \sigma_y \\ \sigma_{xy} \end{bmatrix} = \frac{E}{1 - \nu^2} \begin{bmatrix} 1 & \nu & 0 \\ \nu & 1 & 0 \\ 0 & 0 & \frac{1-\nu}{2} \end{bmatrix} \begin{bmatrix} \varepsilon_x \\ \varepsilon_y \\ 2\varepsilon_{xy} \end{bmatrix}, \tag{5.4}$$

or in matrix notation as

$$\sigma = C\varepsilon, \tag{5.5}$$

where C is the so-called elasticity matrix. It should be noted here that the engineering shear strain $\gamma_{xy} = 2\varepsilon_{xy}$ is used in the formulation of Eq. (5.4).

Rearranging the elastic stiffness form given in Eq. (5.4) for the strains gives the elastic compliance form

$$\begin{bmatrix} \varepsilon_x \\ \varepsilon_y \\ 2\varepsilon_{xy} \end{bmatrix} = \frac{1}{E} \begin{bmatrix} 1 & -\nu & 0 \\ -\nu & 1 & 0 \\ 0 & 0 & 2(\nu + 1) \end{bmatrix} \begin{bmatrix} \sigma_x \\ \sigma_y \\ \sigma_{xy} \end{bmatrix}, \tag{5.6}$$

or in matrix notation as

$$\varepsilon = D\sigma, \tag{5.7}$$

where $D = C^{-1}$ is the so-called elastic compliance matrix. The general characteristic of a plane Hooke's law in the form of Eqs. (5.5) and (5.6) is that two independent material parameters are used.

It should be finally noted that the thickness strain ε_z can be obtained based on the two in-plane normal strains ε_x and ε_y as:

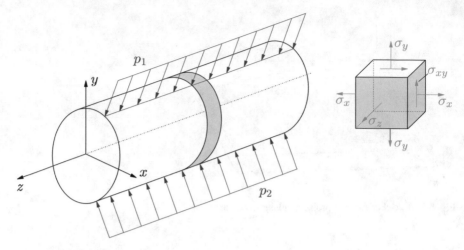

Fig. 5.3 Two dimensional problem: plane strain case

$$\varepsilon_z = -\frac{\nu}{1-\nu} \cdot \left(\varepsilon_x + \varepsilon_y\right). \tag{5.8}$$

The last equation can be derived from the tree-dimensional formulation, see Sect. 8.3.

5.3.2   Plane Strain Case

The two-dimensional plane strain case ($\varepsilon_z = \varepsilon_{yz} = \varepsilon_{xz} = 0$) shown in Fig. 5.3 is commonly used for the analysis of elongated prismatic bodies of uniform cross section subjected to uniform loading along their longitudinal axis but without any component in direction of the z-axis (e. g. pressure p_1 and p_2), such as in the case of tunnels, soil slopes, and retaining walls [3]. It should be noted here that the normal thickness strain is zero ($\varepsilon_z = 0$) whereas the thickness normal stress is present ($\sigma_z \neq 0$).

The plane strain Hooke's law for a linear-elastic isotropic material based on the Young's modulus E and Poisson's ratio ν can be written for a constant temperature as

$$\begin{bmatrix} \sigma_x \\ \sigma_y \\ \sigma_{xy} \end{bmatrix} = \frac{E}{(1+\nu)(1-2\nu)} \begin{bmatrix} 1-\nu & \nu & 0 \\ \nu & 1-\nu & 0 \\ 0 & 0 & \frac{1-2\nu}{2} \end{bmatrix} \cdot \begin{bmatrix} \varepsilon_x \\ \varepsilon_y \\ 2\,\varepsilon_{xy} \end{bmatrix}, \tag{5.9}$$

or in matrix notation as

$$\sigma = C\varepsilon, \tag{5.10}$$

where C is the so-called elasticity matrix.

Rearranging the elastic stiffness form given in Eq. (5.9) for the strains gives the elastic compliance form

$$
\begin{bmatrix} \varepsilon_x \\ \varepsilon_y \\ 2\varepsilon_{xy} \end{bmatrix} = \frac{1-\nu^2}{E} \begin{bmatrix} 1 & -\frac{\nu}{1-\nu} & 0 \\ -\frac{\nu}{1-\nu} & 1 & 0 \\ 0 & 0 & \frac{2}{1-\nu} \end{bmatrix} \begin{bmatrix} \sigma_x \\ \sigma_y \\ \sigma_{xy} \end{bmatrix}, \tag{5.11}
$$

or in matrix notation as

$$
\varepsilon = D\sigma, \tag{5.12}
$$

where $D = C^{-1}$ is the so-called elastic compliance matrix. The general characteristic of a plane strain Hooke's law in the form of Eqs. (5.9) and (5.11) is that two independent material parameters are used.

It should be finally noted that the thickness stress σ_z can be obtained based on the two in-plane normal stresses σ_x and σ_y as:

$$
\sigma_z = \nu(\sigma_x + \sigma_y). \tag{5.13}
$$

The last equation can be derived from the tree-dimensional formulation, see Sect. 8.3

5.4 Equilibrium

Figure 5.4 shows the normal and shear stresses which are acting on a differential volume element in the x-direction. All forces are drawn in their positive direction at each cut face. A positive cut face is obtained if the outward surface normal is directed in the positive direction of the corresponding coordinate axis. This means that the right-hand face in Fig. 5.4 is positive and the force $(\sigma_x + \frac{\partial \sigma_x}{\partial x} dx) dy dz$ is oriented in the positive x-direction. In a similar way, the top face is positive, i.e. the outward surface normal is directed in the positive y-direction, and the shear force[1] is oriented in the positive x-direction. Since the volume element is assumed to be in equilibrium, forces resulting from stresses on the sides of the cuboid and from the body forces f_i $(i = x, y, z)$ must be balanced. These body forces are defined as forces per unit volume which can be produced by gravity,[2] acceleration, magnetic fields, and so on.

The static equilibrium of forces in the x-direction based on the five force components—two normal forces, two shear forces and one body force—indicated in Fig. 5.4 gives

---

[1] In the case of a shear force σ_{ij}, the first index i indicates that the stress acts on a plane normal to the i-axis and the second index j denotes the direction in which the stress acts.

[2] If gravity is acting, the body force f results as the product of density times standard gravity: $f = \frac{F}{V} = \frac{mg}{V} = \frac{m}{V}g = \varrho g$. The units can be checked by consideration of $1\,\text{N} = 1\frac{\text{mkg}}{\text{s}^2}$.

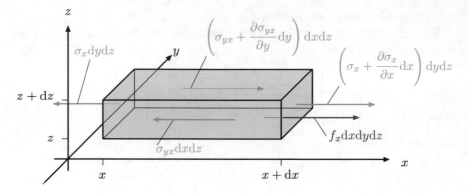

Fig. 5.4 Stress and body forces which act on a plane differential volume element in x-direction (note that the three directions $\mathrm{d}x$, $\mathrm{d}y$ and $\mathrm{d}z$ are differently sketched to indicate the plane problem)

$$\left(\sigma_x + \frac{\partial \sigma_x}{\partial x}\right)\mathrm{d}y\mathrm{d}z - \sigma_x\mathrm{d}y\mathrm{d}z + \left(\frac{\partial \sigma_{yx}}{\partial y}\right)\mathrm{d}x\mathrm{d}z$$

$$- \sigma_{yx}\mathrm{d}x\mathrm{d}z + f_x\mathrm{d}x\mathrm{d}y\mathrm{d}z = 0 , \tag{5.14}$$

or after simplification and canceling with $\mathrm{d}V = \mathrm{d}x\mathrm{d}y\mathrm{d}z$:

$$\frac{\partial \sigma_x}{\partial x} + \frac{\partial \sigma_{yx}}{\partial y} + f_x = 0 . \tag{5.15}$$

Based on the same approach, a similar equation can be specified in the y-direction:

$$\frac{\partial \sigma_y}{\partial y} + \frac{\partial \sigma_{yx}}{\partial x} + f_y = 0 . \tag{5.16}$$

These two balance equations can be written in matrix notation as

$$\begin{bmatrix} \dfrac{\partial}{\partial x} & 0 & \dfrac{\partial}{\partial y} \\[2mm] 0 & \dfrac{\partial}{\partial y} & \dfrac{\partial}{\partial x} \end{bmatrix} \begin{bmatrix} \sigma_x \\ \sigma_y \\ \sigma_{xy} \end{bmatrix} + \begin{bmatrix} f_x \\ f_y \end{bmatrix} = \begin{bmatrix} 0 \\ 0 \end{bmatrix} , \tag{5.17}$$

or in symbolic notation:

$$\mathcal{L}_1^{\mathrm{T}}\sigma + b = 0 , \tag{5.18}$$

where $\mathcal{L}_1$ is the differential operator matrix and b the column matrix of body forces.

Table 5.2 Fundamental governing equations of a continuum in the plane elasticity case

Expression	Matrix notation	Tensor notation
Kinematics	$\varepsilon = \mathcal{L}_1 u$	$\varepsilon_{ij} = \frac{1}{2}\left(u_{i,j} + u_{j,i}\right)$
Constitution	$\sigma = C\varepsilon$	$\sigma_{ij} = C_{ijkl}\varepsilon_{kl}$
Equilibrium	$\mathcal{L}_1^{\mathrm{T}}\sigma + b = 0$	$\sigma_{ij,i} + b_j = 0$

5.5   Differential Equation

The basic equations introduced in the previous three sections, i.e., the kinematics, the constitutive, and the equilibrium equation, are summarized in the following Table 5.2 where in addition the tensor notation[3] is given.

For the solution of the eight unknown spatial functions (2 components of the displacement vector, 3 components of the symmetric strain tensor and 3 components of the symmetric stress tensor), a set of eight scalar field equations is available:

- Equilibrium: 2,
- Constitution: 3,
- Kinematics: 3.

Furthermore, the boundary conditions are given:

$$u \quad \text{on} \quad \Gamma_u, \tag{5.19}$$

$$t \quad \text{on} \quad \Gamma_t, \tag{5.20}$$

where Γ_u is the part of the boundary where a displacement boundary condition is prescribed and Γ_t is the part of the boundary where a traction boundary condition, i.e. external force per unit area, is prescribed with $t_j = \sigma_{ij}n_j$, where n_j are the components of the normal vector.

The eight scalar field equations can be combined to eliminate the stress and strain fields. As a result, two scalar field equations for the three scalar displacement fields are obtained. These equations are called the Lamé-Navier equations and can be derived as follows:

Introducing the constitutive equation according to (5.5) in the equilibrium equation 8.13 gives:

$$\mathcal{L}_1^{\mathrm{T}}C\varepsilon + b = 0. \tag{5.21}$$

Introducing the kinematics relations in the last equation according to (5.3) finally gives the Lamé-Navier equations:

---

[3] A differentiation is there indicated by the use of a comma: The first index refers to the component and the comma indicates the partial derivative with respect to the second subscript corresponding to the relevant coordinate axis, [1].

Table 5.3 Different formulations of the basic equations for plane elasticity (deformation in the x-y plane) E: Young's modulus; ν: Poisson's ratio; f_x volume-specific force in x-direction; f_y volume-specific force in y-direction

Specific formulation	General formulation
Kinematics	
$$\begin{bmatrix} \varepsilon_x \\ \varepsilon_y \\ 2\varepsilon_{xy} \end{bmatrix} = \begin{bmatrix} \frac{\partial}{\partial x} & 0 \\ 0 & \frac{\partial}{\partial y} \\ \frac{\partial}{\partial y} & \frac{\partial}{\partial x} \end{bmatrix} \begin{bmatrix} u_x \\ u_y \end{bmatrix}$$	$\varepsilon = \mathcal{L}_1 u$
Constitution	
$$\begin{bmatrix} \sigma_x \\ \sigma_y \\ \sigma_{xy} \end{bmatrix} = \frac{E'}{1-\nu'^2} \begin{bmatrix} 1 & \nu' & 0 \\ \nu' & 1 & 0 \\ 0 & 0 & \frac{1-\nu'}{2} \end{bmatrix} \begin{bmatrix} \varepsilon_x \\ \varepsilon_y \\ 2\varepsilon_{xy} \end{bmatrix}$$ with $E' = E$ and $\nu' = \nu$ for plane stress and $E' = \dfrac{E}{1-\left(\frac{\nu}{1-\nu}\right)^2}$ and $\nu' = \dfrac{\nu}{1-\nu}$ for plane strain	$\sigma = C\varepsilon$
Equilibrium	
$$\begin{bmatrix} \frac{\partial}{\partial x} & 0 & \frac{\partial}{\partial y} \\ 0 & \frac{\partial}{\partial y} & \frac{\partial}{\partial x} \end{bmatrix} \begin{bmatrix} \sigma_x \\ \sigma_y \\ \sigma_{xy} \end{bmatrix} + \begin{bmatrix} f_x \\ f_y \end{bmatrix} = \begin{bmatrix} 0 \\ 0 \end{bmatrix}$$	$\mathcal{L}_1^T \sigma + b = 0$
PDE	
$$\frac{E'}{1-\nu'^2} \begin{bmatrix} \frac{\partial^2}{\partial x^2} + \frac{1-\nu'}{2}\frac{\partial^2}{\partial y^2} & \nu'\frac{\partial^2}{\partial x\partial y} + \frac{1-\nu'}{2}\frac{\partial^2}{\partial x\partial y} \\ \nu'\frac{\partial^2}{\partial x\partial y} + \frac{1-\nu'}{2}\frac{\partial^2}{\partial x\partial y} & \frac{\partial^2}{\partial y^2} + \frac{1-\nu'}{2}\frac{\partial^2}{\partial x^2} \end{bmatrix}$$ $$\begin{bmatrix} u_x \\ u_y \end{bmatrix} + \begin{bmatrix} f_x \\ f_y \end{bmatrix} = \begin{bmatrix} 0 \\ 0 \end{bmatrix}$$	$\mathcal{L}_1^T C \mathcal{L}_1 u + b = 0$

$$\mathcal{L}_1^T C \mathcal{L}_1 u + b = 0. \tag{5.22}$$

Alternatively, the displacements may be substituted and the differential equations are obtained in terms of stresses. This formulation is known as the Beltrami-Michell equations. If the body forces vanish ($b = 0$), the partial differential equations in terms of stresses are called the Beltrami equations.

Table 5.3 summarizes the different formulations of the basic equations for plane elasticity, once in their specific form and once in symbolic notation.

Table 5.4 shows a comparison between the basic equations for a rod and plane elasticity problem. It can be seen that the use of the differential operator $\mathcal{L}_1\{\ldots\}$ allows to depict a simple analogy between both sets of equations.

Table 5.4 Comparison of basic equations for rod and plane elasticity

Rod	Plane elasticity
Kinematics	
$\varepsilon_x(x) = \mathcal{L}_1(u_x(x))$	$\varepsilon = \mathbf{\mathcal{L}}_1\mathbf{u}$
Constitution	
$\sigma_x(x) = C\varepsilon_x(x)$	$\boldsymbol{\sigma} = \mathbf{C}\boldsymbol{\varepsilon}$
Equilibrium	
$\mathcal{L}_1^{\mathrm{T}}(\sigma_x(x)) + b = 0$	$\mathbf{\mathcal{L}}_1^{\mathrm{T}}\boldsymbol{\sigma} + \mathbf{b} = \mathbf{0}$
PDE	
$\mathcal{L}_1^{\mathrm{T}}(C\mathcal{L}_1(u_x(x))) + b = 0$	$\mathbf{\mathcal{L}}_1^{\mathrm{T}}\mathbf{C}\mathbf{\mathcal{L}}_1\mathbf{u} + \mathbf{b} = \mathbf{0}$

References

1. Chen WF, Han DJ (1988) Plasticity for structural engineers. Springer, New York
2. Eschenauer H, Olhoff N, Schnell W (1997) Applied structural mechanics: fundamentals of elasticity, load-bearing structures, structural optimization. Springer, Berlin
3. Öchsner A (2016) Continuum damage and fracture mechanics. Springer, Singapore

Chapter 6
Classical Plates

Abstract This chapter covers the continuum mechanical description of classical plate members. Classical plates are thin plates where the contribution of the shear force on the deformations is neglected. Based on the three basic equations of continuum mechanics, i.e., the kinematics relationship, the constitutive law, and the equilibrium equation, the partial differential equation, which describes the physical problem, is derived.

6.1 Introduction

A classical plate is defined as a thin structural member, as schematically shown in Fig. 6.1, with a much smaller thickness h than the planar dimensions ($2a$ and $2b$). It can be seen as a two-dimensional extension or generalization of the Euler–Bernoulli beam (see Chap. 3). The following derivations are restricted to some simplifications:

- the thickness h is constant and much smaller than the planer dimensions a and b: $\frac{h}{a}$ and $\frac{h}{b} < 0.1$,
- the thickness h is constant ($\rightarrow \varepsilon_z = 0$) and the undeformed plate shape is planar,
- the displacement $u_z(x, y)$ is small compared to the thickness dimension h: $u_z < 0.2h$,
- the material is isotropic, homogenous and linear-elastic according to Hooke's law for a plane stress state ($\sigma_z = \tau_{xz} = \tau_{yz} = 0$),
- Bernoulli's hypothesis is valid, i.e. a cross-sectional plane stays plane and un-wrapped in the deformed state. This means that the shear strains γ_{yz} and γ_{xz} due to the distributed shear forces q_x and q_y are neglected,
- external forces act only perpendicular to the xy-plane, the vector of external moments lies within the xy-plane, and
- only rectangular plates are considered.

The external loads, which are considered within this chapter, are single forces F_z, single moments M_x and M_y, area distributed forces $q_z(x, y)$, and area distributed moments $m_x(x, y)$ and $m_y(x, y)$.

© The Author(s), under exclusive license to Springer Nature Switzerland AG 2020   57
A. Öchsner, *Partial Differential Equations of Classical Structural Members*,
SpringerBriefs in Continuum Mechanics,
https://doi.org/10.1007/978-3-030-35311-7_6

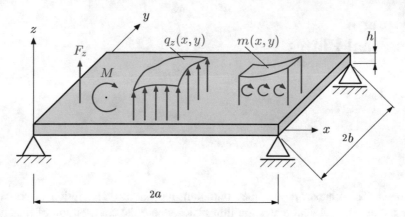

Fig. 6.1 General configuration for a classical plate problem

Table 6.1 Analogies between the classical beam and plate theories

Classical beam	Classical plate
1D	2D
Deformation perpendicular	Deformation perpendicular
To principal x-axis	To xy-plane
u_z, φ_y	$u_z, \varphi_x, \varphi_y$

The classical theories of plate bending distinguish between shear-rigid and shear-flexible models. The shear rigid-plate, also called the classical or Kirchhoff plate, neglects the shear deformation from the shear forces. This theory corresponds to the classical Euler–Bernoulli beam theory (see Chap. 3). The consideration of the shear deformation leads to the Reissner-Mindlin plate (see Chap. 7) which corresponds to the Timoshenko beam (see Chap. 4).

The analogies between the classical beam and plate theories are summarized in Table 6.1.

6.2 Kinematics

The kinematics or strain-displacement relations extract the strain field contained in a displacement field. Let us first derive a kinematics relation which relates the variation of u_x across the plate thickness in terms of the displacement u_z. For this purpose, let us imagine that a plate element is bent around the y-axis, see Fig. 6.2. We assume in the following the same definition of the rotational angle φ_y as in Chap. 3 for the Euler–Bernoulli beam. This means that the angle φ_y is positive if the vector of the rotational direction is pointing in positive y-axis.

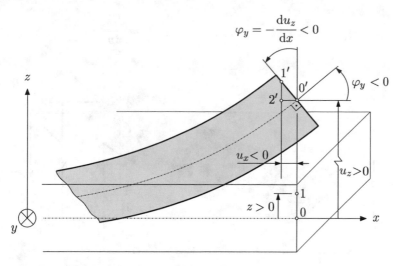

Fig. 6.2 Configuration for the derivation of kinematics relations in the x-z plane. Note that the deformation is exaggerated for better illustration

Looking at the right-angled triangle $0'1'2'$, we can state that[1]

$$\sin(-\varphi_y) = \frac{\overline{2'0'}}{\overline{0'1'}} = \frac{-u_x}{z},$$ (6.1)

which results for small angles $(\sin(-\varphi_y) \approx -\varphi_y)$ in:

$$u_x = +z\varphi_y.$$ (6.2)

Looking at the curved center line in Fig. 6.2, it holds that the slope of the tangent line at $0'$ equals:

$$\tan(-\varphi_y) = +\frac{\mathrm{d}u_z}{\mathrm{d}x} \approx -\varphi_y.$$ (6.3)

If Eqs. (6.2) and (6.3) are combined, the following results:

$$u_x = -z\frac{\mathrm{d}u_z}{\mathrm{d}x}.$$ (6.4)

Considering a plate which is bent around the x-axis (see Fig. 6.3) and following the same line of reasoning (the angle φ_x is assumed positive if the vector of the rotational direction is pointing in positive x-axis.), similar equations can be derived for u_y:

---

[1]Note that according to the assumptions of the classical *thin* plate theory the lengths $\overline{01}$ and $\overline{0'1'}$ remain unchanged.

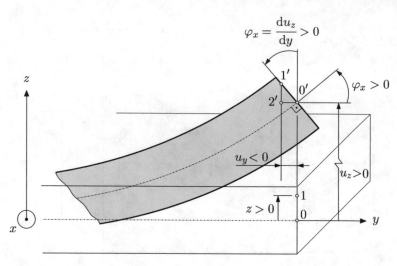

Fig. 6.3 Configuration for the derivation of kinematics relations in the y-z plane. Note that the deformation is exaggerated for better illustration

$$\varphi_x \approx \frac{\mathrm{d}u_z}{\mathrm{d}y}, \tag{6.5}$$

$$u_y = -z\varphi_x, \tag{6.6}$$

$$u_y = -z\frac{\mathrm{d}u_z}{\mathrm{d}y}. \tag{6.7}$$

One may find in the scholarly literature other definitions of the rotational angles [1, 2, 4, 5]. The angle φ_y is introduced in the xz-plane (see Fig. 6.2) whereas φ_x is introduced in the yz-plane (see Fig. 6.3). These definitions are closer to the classical definitions of the angles in the scope of finite elements but not conform with the definitions of the stress resultants (see M_x^{n} and M_y^{n} in Fig. 6.5). Other definitions assume, for example, that the rotational angle φ_x (now defined in the x-z plane) is positive if it leads to a positive displacement u_x at the positive z-side of the neutral axis. The same definition holds for the angle φ_y (now defined in the y-z plane).

Using classical engineering definitions of strain, the following relations can be obtained [1, 3]:

$$\varepsilon_x = \frac{\partial u_x}{\partial x} \overset{(6.4)}{=} \frac{\partial}{\partial x}\left(-z\frac{\partial u_z}{\partial x}\right) = -z\frac{\partial^2 u_z}{\partial x^2} = z\kappa_x, \tag{6.8}$$

$$\varepsilon_y = \frac{\partial u_y}{\partial y} \overset{(6.7)}{=} \frac{\partial}{\partial y}\left(-z\frac{\partial u_z}{\partial y}\right) = -z\frac{\partial^2 u_z}{\partial y^2} = z\kappa_y, \tag{6.9}$$

$$\gamma_{xy} = \frac{\partial u_x}{\partial y} + \frac{\partial u_y}{\partial x} \overset{(6.4),(6.7)}{=} -2z\frac{\partial^2 u_z}{\partial x \partial y} = z\kappa_{xy}. \tag{6.10}$$

In matrix notation, these three relationships can be written as

$$\begin{bmatrix} \varepsilon_x \\ \varepsilon_y \\ \gamma_{xy} \end{bmatrix} = -z \begin{bmatrix} \frac{\partial^2}{\partial x^2} \\ \frac{\partial^2}{\partial y^2} \\ \frac{2\partial^2}{\partial x \partial y} \end{bmatrix} u_z = z \begin{bmatrix} \kappa_x \\ \kappa_y \\ \kappa_{xy} \end{bmatrix}, \tag{6.11}$$

or symbolically as

$$\varepsilon = -z\mathcal{L}_2 u_z = z\kappa. \tag{6.12}$$

Let us recall here that we obtained in Chap. 3 the following kinematics relationship for the Euler–Bernoulli beam, see Eq. (3.16):

$$\varepsilon_x(x,y) = -z\frac{\mathrm{d}^2 u_z(x)}{\mathrm{d}x^2} = z\kappa. \tag{6.13}$$

This last relationship corresponds exactly to Eq. (6.8).

6.3 Constitution

As stated in the introduction of this chapter, the classical plate theory assumes a plane stress state and the constitutive equation can be taken from Sect. 5.3 as:

$$\begin{bmatrix} \sigma_x \\ \sigma_y \\ \sigma_{xy} \end{bmatrix} = \frac{E}{1-\nu^2} \begin{bmatrix} 1 & \nu & 0 \\ \nu & 1 & 0 \\ 0 & 0 & \frac{1-\nu}{2} \end{bmatrix} \begin{bmatrix} \varepsilon_x \\ \varepsilon_y \\ \gamma_{xy} \end{bmatrix}, \tag{6.14}$$

or rearranged for the elastic compliance form:

$$\begin{bmatrix} \varepsilon_x \\ \varepsilon_y \\ \gamma_{xy} \end{bmatrix} = \frac{1}{E} \begin{bmatrix} 1 & -\nu & 0 \\ -\nu & 1 & 0 \\ 0 & 0 & 2(\nu+1) \end{bmatrix} \begin{bmatrix} \sigma_x \\ \sigma_y \\ \sigma_{xy} \end{bmatrix}. \tag{6.15}$$

The last two equations can be written in matrix notation as

$$\sigma = C\varepsilon, \tag{6.16}$$

or

$$\varepsilon = D\sigma, \tag{6.17}$$

Fig. 6.4 Stresses acting on a classical plate element

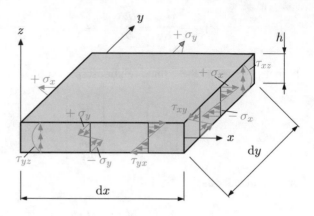

where $\boldsymbol{C}$ is the elasticity matrix and $\boldsymbol{D} = \boldsymbol{C}^{-1}$ is the elastic compliance matrix.

Let us recall here that in Chap. 3 we obtained the following constitutive relationship for the Euler–Bernoulli beam, see Eq. (3.20):

$$\sigma_x = E\varepsilon_x . \tag{6.18}$$

This last relationship corresponds to Eq. (6.14).

6.4   Equilibrium

Let us first look at the stress distributions through the thickness of a classical plate element $\mathrm{d}x\mathrm{d}yh$ as shown in Fig. 6.4. Linear distributed normal stresses (σ_x, σ_y), linear distributed shear stresses (τ_{yx}, τ_{xy}), and parabolic distributed shear stresses (τ_{yz}, τ_{xz}) can be identified. These stresses can be expressed by the so-called stress resultants, i.e. bending moments and shear forces as shown in Fig. 6.5. These stress resultants are taken to be positive if they cause a tensile stress (positive) at a point with positive z-coordinate.

These stress resultants are obtained as in the case of beams[2] by integrating over the stress distributions. In the case of plates, however, the integration is only performed over the thickness, i.e. the moments and forces are given per unit length (normalized with the corresponding side length of the plate element). The normalized (superscript 'n') bending moments are obtained as:

---

[2]See Eq. (3.23) for the Eurler-Bernoulli beam or Eq. (4.11) for the Timoshenko beam.

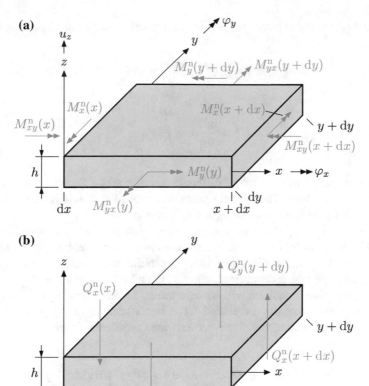

Fig. 6.5 Stress resultants acting on a classical plate element: **a** bending and twisting moments and **b** shear forces. Positive directions are drawn

$$M_x^n = \frac{M_x}{dy} = \int\limits_{-h/2}^{h/2} z\sigma_x dz \, , \tag{6.19}$$

$$M_y^n = \frac{M_y}{dx} = \int\limits_{-h/2}^{h/2} z\sigma_y dz \, . \tag{6.20}$$

The twisting moment per unit length reads:

$$M_{xy}^n = M_{yx}^n = \frac{M_{xy}}{dy} = \frac{M_{yx}}{dx} = \int\limits_{-h/2}^{h/2} z\tau_{xy} dz \, . \tag{6.21}$$

Furthermore, the shear forces per unit length are calculated in the following way:

$$Q_x^n = \frac{Q_x}{dy} = \int_{-h/2}^{h/2} \tau_{xz} dz \, , \tag{6.22}$$

$$Q_y^n = \frac{Q_y}{dx} = \int_{-h/2}^{h/2} \tau_{yz} dz \, . \tag{6.23}$$

It should be noted that a slightly different notation when compared to the beam problems is used here. The bending moment around the y-axis is now called M_x^n (which directly corresponds to the causing stress σ_x) while in the beam notation it was M_y, see Fig. 3.12. Nevertheless, the orientation remains the same. The shear force, which was in the case of the beams given as Q_z is now either Q_x^n or Q_y^n. Thus, in the case of this plate notation, the index refers rather to the plane (check the surface normal vector) in which the corresponding resultant (vector) is located.

The equilibrium condition will be determined in the following for the vertical forces. Assuming that the distributed force is constant ($q_z(x, y) \rightarrow q_z$) and that forces in the direction of the positive z-axis are considered positive, the following results:

$$- Q_x^n(x)dy - Q_y^n(y)dx + Q_x^n(x + dx)dy + Q_y^n(y + dy)dx + q_z dxdy = 0 \, . \tag{6.24}$$

Evaluating the shear forces at $x + dx$ and $y + dy$ in a Taylor's series of first order, meaning

$$Q_x^n(x + dx) \approx Q_x^n(x) + \frac{\partial Q_x^n}{\partial x} dx \, , \tag{6.25}$$

$$Q_y^n(y + dy) \approx Q_y^n(y) + \frac{\partial Q_y^n}{\partial y} dy \, , \tag{6.26}$$

Equation (6.24) results in

$$\frac{\partial Q_x^n}{\partial x} dxdy + \frac{\partial Q_y^n}{\partial y} dydx + q_z dxdy = 0 \, , \tag{6.27}$$

or alternatively after simplification to:

$$\frac{\partial Q_x^n}{\partial x} + \frac{\partial Q_y^n}{\partial y} + q_z = 0 \, . \tag{6.28}$$

The equilibrium of moments around the reference axis at $x + dx$ (positive if the moment vector is pointing in positive y-axis) gives:

$$M_x^n(x + dx)dy - M_x^n(x)dy + M_{yx}^n(y + dy)dx - M_{yx}^n dx$$
$$- Q_y^n(y)dx\frac{dx}{2} + Q_y^n(y + dy)dx\frac{dx}{2} - Q_x^n(x)dydx + q_z dxdy\frac{dx}{2} = 0. \quad (6.29)$$

Expanding the stress resultants at $x + dx$ and $y + dy$ into a Taylor's series of first order, meaning

$$M_x^n(x + dx) = M_x^n(x) + \frac{\partial M_x^n}{\partial x}dx , \quad (6.30)$$

$$M_{yx}^n(y + dy) = M_{yx}^n(y) + \frac{\partial M_{yx}^n}{\partial y}dy , \quad (6.31)$$

$$Q_y^n(y + dy) = Q_y^n(y) + \frac{\partial Q_y^n}{\partial y}dy , \quad (6.32)$$

Equation (6.29) results in

$$\frac{\partial M_x^n}{\partial x}dxdy + \frac{\partial M_{yx}^n}{\partial y}dydx + \frac{\partial Q_y^n}{\partial y}dydx\frac{dx}{2} - Q_x^n(x)dydx + q_z dxdy\frac{dx}{2} = 0. \quad (6.33)$$

Seeing that the terms of third order $(dxdydx)$ are considered as infinitesimally small and because of $M_{yx}^n = M_{xy}^n$, finally the following results:

$$\frac{\partial M_x^n}{\partial x} + \frac{\partial M_{xy}^n}{\partial y} - Q_x^n = 0. \quad (6.34)$$

In a similar way, the equilibrium of moments around the reference axis at $y + dy$ finally gives:

$$\frac{\partial M_y^n}{\partial y} + \frac{\partial M_{xy}^n}{\partial x} - Q_y^n = 0. \quad (6.35)$$

Thus, the three equilibrium equations can be summarized as follows:

$$\frac{\partial Q_x^n}{\partial x} + \frac{\partial Q_y^n}{\partial y} + q_z = 0, \quad (6.36)$$

$$\frac{\partial M_x^n}{\partial x} + \frac{\partial M_{xy}^n}{\partial y} - Q_x^n = 0, \quad (6.37)$$

$$\frac{\partial M_y^n}{\partial y} + \frac{\partial M_{xy}^n}{\partial x} - Q_y^n = 0. \quad (6.38)$$

Let us recall here that we obtained in Chap. 3 the following equilibrium equations for the Euler–Bernoulli beam, see Eqs. (3.35) and (3.36):

$$\frac{\mathrm{d}M_y(x)}{\mathrm{d}x} = Q_z(x)\,, \quad \frac{\mathrm{d}^2 M_y(x)}{\mathrm{d}x^2} = \frac{\mathrm{d}Q_z(x)}{\mathrm{d}x} = -q_z\,. \tag{6.39}$$

Rearranging Eqs. (6.37) and (6.38) for Q^n and introducing in Eq. (6.36) finally gives the combined equilibrium equation as:

$$\frac{\partial^2 M_x^n}{\partial x^2} + 2\frac{\partial^2 M_{xy}}{\partial x \partial y} + \frac{\partial^2 M_y^n}{\partial y^2} + q_z = 0\,. \tag{6.40}$$

The last equation can be written in matrix notation as

$$\left[\frac{\partial^2}{\partial x^2} \; \frac{\partial^2}{\partial y^2} \; \frac{2\partial^2}{\partial x \partial y} \right] \begin{bmatrix} M_x^n \\ M_y^n \\ M_{xy}^n \end{bmatrix} + q_z = 0\,, \tag{6.41}$$

or symbolically as

$$\mathcal{L}_2^{\mathrm{T}} M^n + q_z = 0\,. \tag{6.42}$$

Equations (6.37) and (6.38) can be rearranged to obtain a relationship between the moments and shear forces similar to Eq. (6.39)$_1$:

$$\mathcal{L}_1^{\mathrm{T}} M^n = Q^n\,, \tag{6.43}$$

where the first-order differential operator matrix $\mathcal{L}_1$ is given by Eqs. (5.17) and (5.18).

6.5   Differential Equation

Let us combine the three equations for the resulting moments according to Eqs. (6.19)–(6.21) in matrix notation as

$$M^n = \begin{bmatrix} M_x^n \\ M_y^n \\ M_{xy}^n \end{bmatrix} = \int_{-h/2}^{h/2} z \begin{bmatrix} \sigma_x \\ \sigma_y \\ \tau_{xy} \end{bmatrix} \mathrm{d}z = \int_{-h/2}^{h/2} z\boldsymbol{\sigma}\mathrm{d}z\,. \tag{6.44}$$

Introducing Hooke's law (6.16) and the kinematics relation (6.12) gives for a constant elasticity matrix C

$$M^n = -\int_{-h/2}^{h/2} z^2 C\mathcal{L}_2 u_z \mathrm{d}z = -C\mathcal{L}_2 u_z \underbrace{\int_{-h/2}^{h/2} z^2 \mathrm{d}z}_{\frac{h^3}{12}} = -\underbrace{\frac{h^3}{12}C}_{D}\mathcal{L}_2 u_z\,, \tag{6.45}$$

where the plate elasticity matrix D is given by

$$D = \frac{h^3}{12}C = \underbrace{\frac{Eh^3}{12(1-\nu^2)}}_{D}\begin{bmatrix} 1 & \nu & 0 \\ \nu & 1 & 0 \\ 0 & 0 & \frac{1-\nu}{2} \end{bmatrix},\qquad(6.46)$$

and $D = \frac{Eh^3}{12(1-\nu^2)}$ is the bending rigidity of the plate. Using the kinematics relation in the curvature form (see Eq. (6.12)), it can be stated that

$$M^{\mathrm{n}} = D\kappa.\qquad(6.47)$$

Introducing the moment-displacement relation (6.45) in the equilibrium equation (6.42) results in the plate bending differential equation in the form:

Table 6.2 Different formulations of the basic equations for a classical plate (bending perpendicular to the x-y plane). E: Young's modulus; ν: Poisson's ratio; q_z: area-specific distributed force; h plate thickness; M^{n}: length-specific moment; Q^{n}: length-specific shear force

Specific formulation	General formulation
Kinematics	
$\begin{bmatrix} \varepsilon_x \\ \varepsilon_y \\ \gamma_{xy} \end{bmatrix} = -z \begin{bmatrix} \frac{\partial^2}{\partial x^2} \\ \frac{\partial^2}{\partial y^2} \\ \frac{2\partial^2}{\partial x\partial y} \end{bmatrix} u_z = z \begin{bmatrix} \kappa_x \\ \kappa_y \\ \kappa_{xy} \end{bmatrix}$	$\varepsilon(x,y,z) = -z\mathcal{L}_2 u_z = z\kappa$
Constitution	
$\begin{bmatrix} \sigma_x \\ \sigma_y \\ \sigma_{xy} \end{bmatrix} = \frac{E}{1-\nu^2} \begin{bmatrix} 1 & \nu & 0 \\ \nu & 1 & 0 \\ 0 & 0 & \frac{1-\nu}{2} \end{bmatrix} \begin{bmatrix} \varepsilon_x \\ \varepsilon_y \\ \gamma_{xy} \end{bmatrix}$	$\sigma = C\varepsilon$
$\begin{bmatrix} M_x^{\mathrm{n}} \\ M_y^{\mathrm{n}} \\ M_{xy}^{\mathrm{n}} \end{bmatrix} = \frac{Eh^3}{12(1-\nu^2)} \begin{bmatrix} 1 & \nu & 0 \\ \nu & 1 & 0 \\ 0 & 0 & \frac{1-\nu}{2} \end{bmatrix} \begin{bmatrix} \kappa_x \\ \kappa_y \\ \kappa_{xy} \end{bmatrix}$	$M^{\mathrm{n}} = D\kappa$
Equilibrium	
$\frac{\partial^2 M_x^{\mathrm{n}}}{\partial x^2} + 2\frac{\partial^2 M_{xy}^{\mathrm{n}}}{\partial x\partial y} + \frac{\partial^2 M_y^{\mathrm{n}}}{\partial y^2} + q_z = 0$	$\mathcal{L}_2^{\mathrm{T}} M^{\mathrm{n}} + q_z = 0$
$\begin{bmatrix} \frac{\partial}{\partial x} & 0 & \frac{\partial}{\partial y} \\ 0 & \frac{\partial}{\partial y} & \frac{\partial}{\partial x} \end{bmatrix} \begin{bmatrix} M_x^{\mathrm{n}} \\ M_y^{\mathrm{n}} \\ M_{xy}^{\mathrm{n}} \end{bmatrix} = \begin{bmatrix} Q_x^{\mathrm{n}} \\ Q_y^{\mathrm{n}} \end{bmatrix}$	$\mathcal{L}_1^{\mathrm{T}} M^{\mathrm{n}} = Q^{\mathrm{n}}$
PDE	
$\frac{Eh^3}{12(1-\nu^2)}\left(\frac{\partial^4 u_z}{\partial x^4} + 2\frac{\partial^4 u_z}{\partial x^2\partial y^2} + \frac{\partial^4 u_z}{\partial y^4}\right) = q_z$	$\mathcal{L}_2^{\mathrm{T}}\left(D\mathcal{L}_2 u_z\right) - q_z = 0$

Table 6.3 Comparison of basic equations for an Euler–Bernoulli beam and a Kirchhoff plate (bending in z-direction)

Euler–Bernoulli beam	Kirchhoff plate
Kinematics	
$\varepsilon_x(x,z) = -z\mathcal{L}_2(u_z(x))$	$\varepsilon(x,y,z) = -z\mathcal{L}_2 u_z(x,y)$
$\kappa(x) = -\mathcal{L}u_z(x)$	$\kappa(x,y) = -\mathcal{L}_2 u_z(x,y)$
Constitution	
$\sigma_x(x,z) = C\varepsilon_x(x,z)$	$\sigma(x,y,z) = C\varepsilon(x,y,z)$
$M_y(x) = D\kappa(x)$	$M^n(x,y) = D\kappa(x,y)$
Equilibrium	
$\mathcal{L}_2^T\left(M_y(x)\right) + q_z(x) = 0$	$\mathcal{L}_2^T M^n(x,y) + q_z(x,y) = 0$
PDE	
$\mathcal{L}_2^T\left(D\mathcal{L}_2\left(u_z(x)\right)\right) - q_z(x) = 0$	$\mathcal{L}_2^T\left(D\mathcal{L}_2 u_z(x,y)\right) - q_z(x,y) = 0$

$$\mathcal{L}_2^T\left(D\mathcal{L}_2 u_z\right) - q_z = 0 \,. \tag{6.48}$$

Using the definitions for $\mathcal{L}_2$ and D given in Eqs. (6.41) and (6.46), the following classical form of the plate bending differential equation can be obtained:

$$\frac{Eh^3}{12(1-\nu^2)}\left(\frac{\partial^4 u_z}{\partial x^4} + 2\frac{\partial^4 u_z}{\partial x^2 \partial y^2} + \frac{\partial^4 u_z}{\partial y^4}\right) = q_z \,. \tag{6.49}$$

Let us recall here that in Chap. 3 we obtained the following partial differential equation for the Euler–Bernoulli beam, see Table 3.5:

$$EI_y\frac{d^4 u_z(x)}{dx^4} = q_z(x) \,. \tag{6.50}$$

Table 6.2 summarizes the different formulations of the basic equations for a classical plate and Table 6.3 compares the general formulations with the relations for the Euler–Bernoulli beam.

References

1. Blaauwendraad J (2010) Plates and FEM: surprises and pitfalls. Springer, Dordrecht
2. Reddy JN (2006) An introduction to the finite element method. McGraw Hill, Singapore
3. Timoshenko S, Woinowsky-Krieger S (1959) Theory of plates and shells. McGraw-Hill Book Company, New York
4. Ventsel E, Krauthammer T (2001) Thin plates and shells: theory, analysis, and applications. Marcel Dekker, New York
5. Wang CM, Reddy JN, Lee KH (2000) Shear deformable beams and plates: relationships with classical solution. Elsevier, Oxford

Chapter 7
Shear Deformable Plates

Abstract This chapter covers the continuum mechanical description of thick plate members. Thick plates are plates where the contribution of the shear force on the deformations is considered. Based on the three basic equations of continuum mechanics, i.e., the kinematics relationship, the constitutive law, and the equilibrium equation, the partial differential equations, which describes the physical problem, is derived.

7.1 Introduction

A thick plate is similarly defined as a thin plate (see Fig. 6.1). However, the condition that the thickness h is *much* smaller than the planar dimensions is weakened. The thickness is still smaller than a and b and not in the same range. The case of $h \approx a \approx b$ would rather refer to a three-dimensional element (see Chap. 8). The thick plate can be seen as a two-dimensional extension or generalization of the Timoshenko beam and is also called the Reissner–Mindlin plate[1] in the finite element context.

7.2 Kinematics

Following the procedure outlined in Sect. 6.2, the relationships between the in-plane displacements and rotational angles are, see Fig. 7.1:

$$u_x = +z\phi_y \; ; \;\; u_y = -z\phi_x . \tag{7.1}$$

Expanding the classical relationships for a plane stress state as given in Eqs. (6.8)–(6.10) by two through-thickness shear strains, the following five relations can be given:

---

[1] Strictly speaking, there is a small difference between the plate theory according to Reissner [4] and Mindlin [2] and only for zero Poisson's ratio both derivations are the same.

© The Author(s), under exclusive license to Springer Nature Switzerland AG 2020
A. Öchsner, *Partial Differential Equations of Classical Structural Members*,
SpringerBriefs in Continuum Mechanics,
https://doi.org/10.1007/978-3-030-35311-7_7

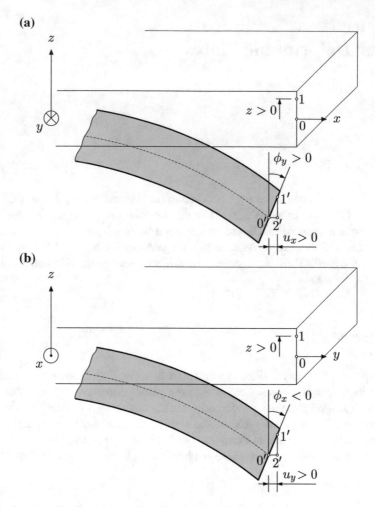

Fig. 7.1 Configuration for the derivation of kinematics relations: **a** xz-plane and **b** yz-plane. Note that the deformation is exaggerated for better illustration

$$\varepsilon_x = \frac{\partial u_x}{\partial x} \; ; \; \varepsilon_y = \frac{\partial u_y}{\partial y} \; ; \; \gamma_{xy} = \frac{\partial u_x}{\partial y} + \frac{\partial u_y}{\partial x} \; ; \tag{7.2}$$

$$\gamma_{xz} = \frac{\partial u_x}{\partial z} + \frac{\partial u_z}{\partial x} \; ; \; \gamma_{yz} = \frac{\partial u_y}{\partial z} + \frac{\partial u_z}{\partial y} . \tag{7.3}$$

Considering the results from Eq. (7.1), the five kinematics relationships can be specialized to:

$$\varepsilon_x = z\frac{\partial \phi_y}{\partial x} = z\kappa_x \; ; \; \varepsilon_y = -z\frac{\partial \phi_x}{\partial y} = z\kappa_y \; ; \; \gamma_{xy} = z\left(\frac{\partial \phi_y}{\partial y} - \frac{\partial \phi_x}{\partial x}\right) = z\kappa_{xy} \; ;$$

$$\gamma_{xz} = \phi_y + \frac{\partial u_z}{\partial x} \; ; \; \gamma_{yz} = -\phi_x + \frac{\partial u_z}{\partial y} . \tag{7.4}$$

In matrix notation, these three relationships can be written as

$$
\begin{bmatrix}
\dfrac{\partial \phi_y}{\partial x} \\[2mm]
-\dfrac{\partial \phi_x}{\partial y} \\[2mm]
\dfrac{\partial \phi_y}{\partial y} - \dfrac{\partial \phi_x}{\partial x} \\[2mm]
\phi_y + \dfrac{\partial u_z}{\partial x} \\[2mm]
-\phi_x + \dfrac{\partial u_z}{\partial y}
\end{bmatrix}
=
\begin{bmatrix}
0 & 0 & \dfrac{\partial}{\partial x} \\[2mm]
0 & -\dfrac{\partial}{\partial y} & 0 \\[2mm]
0 & -\dfrac{\partial}{\partial x} & \dfrac{\partial}{\partial y} \\[2mm]
\dfrac{\partial}{\partial x} & 0 & 1 \\[2mm]
\dfrac{\partial}{\partial y} & -1 & 0
\end{bmatrix}
\begin{bmatrix}
u_z \\[1mm] \phi_x \\[1mm] \phi_y
\end{bmatrix} , \tag{7.5}
$$

or symbolically as

$$\boldsymbol{e} = \mathcal{L}_1 \boldsymbol{u} . \tag{7.6}$$

One may find in the scholarly literature other definitions of the rotational angles [1, 3, 5, 6]. The angle ϕ_y is introduced in the xz-plane (see Fig. 7.1a) whereas ϕ_x is introduced in the yz-plane (see Fig. 7.1b). These definitions are closer to the classical definitions of the angles in the scope of finite elements but not conform with the definitions of the stress resultants (see M_x^n and M_y^n in Fig. 7.2). Other definitions assume, for example, that the rotational angle φ_x (now defined in the x-z plane) is positive if it leads to a positive displacement u_x at the positive z-side of the neutral axis. The same definition holds for the angle φ_y (now defined in the y-z plane).

7.3 Constitution

Let us start to assemble the constitutive equation based on the plane stress formulation for a *thin* plate as given in Table 6.2:

$$\begin{bmatrix} M_x^n \\ M_y^n \\ M_{xy}^n \end{bmatrix} = \underbrace{\frac{Eh^3}{12(1-\nu^2)} \begin{bmatrix} 1 & \nu & 0 \\ \nu & 1 & 0 \\ 0 & 0 & \frac{1-\nu}{2} \end{bmatrix}}_{D_b} \begin{bmatrix} \kappa_x \\ \kappa_y \\ \kappa_{xy} \end{bmatrix}, \tag{7.7}$$

$$\underbrace{\phantom{\frac{Eh^3}{12(1-\nu^2)}}}_{D_b}$$

or under consideration of the generalized strains e as (see Eq. (7.5)):

$$\begin{bmatrix} M_x^n \\ \\ M_y^n \\ \\ M_{xy}^n \end{bmatrix} = \frac{Eh^3}{12(1-\nu^2)} \begin{bmatrix} 1 & \nu & 0 \\ \nu & 1 & 0 \\ 0 & 0 & \frac{1-\nu}{2} \end{bmatrix} \begin{bmatrix} \dfrac{\partial \phi_y}{\partial x} \\ -\dfrac{\partial \phi_x}{\partial y} \\ \dfrac{\partial \phi_y}{\partial y} - \dfrac{\partial \phi_x}{\partial x} \end{bmatrix}. \tag{7.8}$$

In extension to the equations for the Timoshenko beam (see Eqs. 4.22 and (4.13)), the two through-thickness shear strains can be related to the normalized shear forces by:

$$\begin{bmatrix} -Q_x^n \\ -Q_y^n \end{bmatrix} = -k_s Gh \underbrace{\begin{bmatrix} \gamma_{xz} \\ \gamma_{yz} \end{bmatrix}}_{D_s} = \underbrace{-k_s Gh \begin{bmatrix} 1 & 0 \\ 0 & 1 \end{bmatrix}}_{D_s} \begin{bmatrix} \gamma_{xz} \\ \gamma_{yz} \end{bmatrix}, \tag{7.9}$$

where the minus sign was only introduced for formal reasons to have a certain consistency in the further derivations (see also the constitutive equation for the Timoshenko beam in Table 4.5).

Both equations for the constitutive contributions (see Eqs. (7.8) and (7.9)) can be combined to a single matrix form:

$$\begin{bmatrix} M_x^n \\ M_y^n \\ M_{xy}^n \\ -Q_x^n \\ -Q_y^n \end{bmatrix} = \begin{bmatrix} \dfrac{Eh^3}{12(1-\nu^2)} \begin{bmatrix} 1 & \nu & 0 \\ \nu & 1 & 0 \\ 0 & 0 & \frac{1-\nu}{2} \end{bmatrix} & \begin{bmatrix} 0 & 0 \\ 0 & 0 \\ 0 & 0 \end{bmatrix} \\ \begin{bmatrix} 0 & 0 & 0 \\ 0 & 0 & 0 \end{bmatrix} & -k_s Gh \begin{bmatrix} 1 & 0 \\ 0 & 1 \end{bmatrix} \end{bmatrix} \begin{bmatrix} \dfrac{\partial \phi_y}{\partial x} \\ -\dfrac{\partial \phi_x}{\partial y} \\ \dfrac{\partial \phi_y}{\partial y} - \dfrac{\partial \phi_x}{\partial x} \\ \gamma_{xz} \\ \gamma_{yz} \end{bmatrix}, \tag{7.10}$$

or symbolically as

$$s = De, \tag{7.11}$$

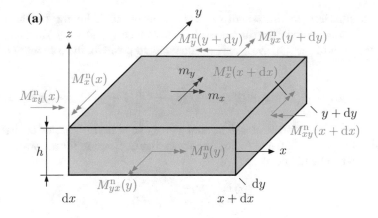

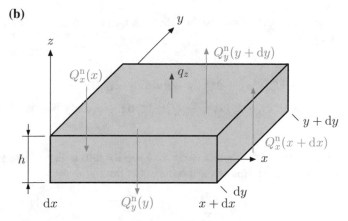

Fig. 7.2 Stress resultants acting on a thick plate element: **a** bending and twisting moments (the distributed moments m_i have the same positive direction as the rotational angles ϕ_i, see Fig. 7.1) and **b** shear forces. Positive directions are drawn

where $\boldsymbol{D}$ is the plate elasticity matrix.[2]

7.4 Equilibrium

The derivation of the equilibrium equations follows the line of reasoning which was introduced in Sect. 6.4 for thin plates. In addition, we consider in the following the area distributed moments $m_x(x, y)$ and $m_y(x, y)$, see Fig. 7.2.

---

[2] This plate elasticity matrix should not be confused with the compliance matrix which is represented by the same symbol.

The equilibrium condition will be determined in the following for the vertical forces. Assuming that the distributed load is constant ($q_z(x, y) \rightarrow q_z$) and that forces in the direction of the positive z-axis are considered positive, the following results:

$$- Q_x^n(x)\mathrm{d}y - Q_y^n(y)\mathrm{d}x + Q_x^n(x + \mathrm{d}x)\mathrm{d}y + Q_y^n(y + \mathrm{d}y)\mathrm{d}x + q_z\mathrm{d}x\mathrm{d}y = 0.$$
(7.12)

Evaluating the shear forces at $x + \mathrm{d}x$ and $y + \mathrm{d}y$ in a Taylor's series of first order as outlined in Eqs. (6.25) and (6.26), the following expression for the vertical force equilibrium can be obtained:

$$\frac{\partial Q_x^n}{\partial x} + \frac{\partial Q_y^n}{\partial y} + q_z = 0.$$
(7.13)

The equilibrium of moments around the reference axis at $x + \mathrm{d}x$ (positive if parallel to the y-axis) gives:

$$M_x^n(x + \mathrm{d}x)\mathrm{d}y - M_x^n(x)\mathrm{d}y + M_{yx}^n(y + \mathrm{d}y)\mathrm{d}x - M_{yx}^n\mathrm{d}x$$
$$- Q_y^n(y)\mathrm{d}x\tfrac{\mathrm{d}x}{2} + Q_y^n(y + \mathrm{d}y)\mathrm{d}x\tfrac{\mathrm{d}x}{2} - Q_x^n(x)\mathrm{d}y\mathrm{d}x + q_z\mathrm{d}x\mathrm{d}y\tfrac{\mathrm{d}x}{2} + m_y\mathrm{d}x\mathrm{d}y = 0.$$
(7.14)

Expanding the stress resultants at $x + \mathrm{d}x$ and $y + \mathrm{d}y$ into a Taylor's series of first order and neglecting the terms of third order gives finally:

$$\frac{\partial M_x^n}{\partial x} + \frac{\partial M_{xy}^n}{\partial y} - Q_x^n + m_y = 0.$$
(7.15)

In a similar way, we can write the moment equilibrium around the x-axis (with the reference axis at $y + \mathrm{d}y$):

$$- M_y^n(y + \mathrm{d}y)\mathrm{d}x + M_y^n(y)\mathrm{d}x - M_{xy}^n(x + \mathrm{d}x)\mathrm{d}y + M_{xy}^n(x)\mathrm{d}y$$
$$- Q_x^n(x + \mathrm{d}x)\mathrm{d}y\tfrac{\mathrm{d}y}{2} + Q_x^n(x)\mathrm{d}y\tfrac{\mathrm{d}y}{2} + Q_y^n(y)\mathrm{d}x\mathrm{d}y + m_x\mathrm{d}x\mathrm{d}y = 0.$$
(7.16)

Expanding the stress resultants at $x + \mathrm{d}x$ and $y + \mathrm{d}y$ into a Taylor's series of first order and neglecting the terms of third order gives finally:

$$\frac{\partial M_y^n}{\partial y} + \frac{\partial M_{xy}^n}{\partial x} - Q_y^n - m_x = 0.$$
(7.17)

The three equilibrium equations (see Eqs. (7.13), (7.15) and (7.17)) can be written in matrix notation as

$$
\begin{bmatrix}
0 & 0 & 0 & \dfrac{\partial}{\partial x} & \dfrac{\partial}{\partial y} \\[2mm]
0 & -\dfrac{\partial}{\partial y} & -\dfrac{\partial}{\partial x} & 0 & -1 \\[2mm]
\dfrac{\partial}{\partial x} & 0 & \dfrac{\partial}{\partial y} & 1 & 0
\end{bmatrix}
\begin{bmatrix}
M_x^n \\[1mm] M_y^n \\[1mm] M_{xy}^n \\[1mm] -Q_x^n \\[1mm] -Q_y^n
\end{bmatrix}
+
\begin{bmatrix}
-q_z \\ m_x \\ m_y
\end{bmatrix}
=
\begin{bmatrix}
0 \\ 0 \\ 0
\end{bmatrix},
\tag{7.18}
$$

or symbolically as

$$
\mathcal{L}_1^{\mathrm{T}} s + b = 0.
\tag{7.19}
$$

7.5 Differential Equation

Introducing the constitutive equation (7.11) and the kinematics equation (7.6) in the equilibrium equation (7.19) gives the general rule for the derivation of the differential equation as:

$$
\mathcal{L}_1^{\mathrm{T}} D \mathcal{L}_1 u + b = 0.
\tag{7.20}
$$

The first matrix multiplication, i.e. $\mathcal{L}_1^{\mathrm{T}} D$, reads as:

$$
\begin{bmatrix}
0 & 0 & 0 & \dfrac{\partial}{\partial x} & \dfrac{\partial}{\partial y} \\[2mm]
0 & -\dfrac{\partial}{\partial y} & -\dfrac{\partial}{\partial x} & 0 & -1 \\[2mm]
\dfrac{\partial}{\partial x} & 0 & \dfrac{\partial}{\partial y} & 1 & 0
\end{bmatrix}
\left[
\begin{array}{c|c}
D_b \begin{bmatrix} 1 & \nu & 0 \\ \nu & 1 & 0 \\ 0 & 0 & \frac{1-\nu}{2} \end{bmatrix} & \begin{bmatrix} 0 & 0 \\ 0 & 0 \\ 0 & 0 \end{bmatrix} \\[5mm]
\hline
\begin{bmatrix} 0 & 0 & 0 \\ 0 & 0 & 0 \end{bmatrix} & -D_s \begin{bmatrix} 1 & 0 \\ 0 & 1 \end{bmatrix}
\end{array}
\right]
=
$$

$$
\begin{bmatrix}
0 & 0 & 0 & -D_s\dfrac{\partial}{\partial x} & -D_s\dfrac{\partial}{\partial y} \\[3mm]
-D_b\dfrac{\partial}{\partial y}\nu & -D_b\dfrac{\partial}{\partial y} & -D_b\dfrac{\partial}{\partial x}\dfrac{1-\nu}{2} & 0 & +D_s \\[3mm]
D_b\dfrac{\partial}{\partial x} & D_b\dfrac{\partial}{\partial x}\nu & D_b\dfrac{\partial}{\partial y}\dfrac{1-\nu}{2} & -D_s & 0
\end{bmatrix}.
\tag{7.21}
$$

The second matrix multiplication, i.e. $(\mathcal{L}_1^{\mathrm{T}} D)\mathcal{L}_1$, reads

$$
\begin{bmatrix}
0 & 0 & 0 & -D_s\dfrac{\partial}{\partial x} & -D_s\dfrac{\partial}{\partial y} \\[2mm]
-D_b\dfrac{\partial}{\partial y}\nu & -D_b\dfrac{\partial}{\partial y} & -D_b\dfrac{\partial}{\partial x}\dfrac{1-\nu}{2} & 0 & +D_s \\[2mm]
D_b\dfrac{\partial}{\partial x} & D_b\dfrac{\partial}{\partial x}\nu & D_b\dfrac{\partial}{\partial y}\dfrac{1-\nu}{2} & -D_s & 0
\end{bmatrix}
\begin{bmatrix}
0 & 0 & \dfrac{\partial}{\partial x} \\[2mm]
0 & -\dfrac{\partial}{\partial y} & 0 \\[2mm]
0 & -\dfrac{\partial}{\partial x} & \dfrac{\partial}{\partial y} \\[2mm]
\dfrac{\partial}{\partial x} & 0 & 1 \\[2mm]
\dfrac{\partial}{\partial y} & -1 & 0
\end{bmatrix},
$$

$$(7.22)$$

which finally results in the following matrix form of the differential equation:

$$
\begin{bmatrix}
-D_s\left(\dfrac{\partial^2}{\partial x^2}+\dfrac{\partial^2}{\partial y^2}\right) & D_s\dfrac{\partial}{\partial y} & -D_s\dfrac{\partial}{\partial x} \\[3mm]
D_s\dfrac{\partial}{\partial y} & D_b\left(\dfrac{1-\nu}{2}\dfrac{\partial^2}{\partial x^2}+\dfrac{\partial^2}{\partial y^2}\right)-D_s & -\dfrac{1+\nu}{2}D_b\dfrac{\partial^2}{\partial x\partial y} \\[3mm]
-D_s\dfrac{\partial}{\partial x} & -\dfrac{1+\nu}{2}D_b\dfrac{\partial^2}{\partial x\partial y} & D_b\left(\dfrac{\partial^2}{\partial x^2}+\dfrac{1-\nu}{2}\dfrac{\partial^2}{\partial y^2}\right)-D_s
\end{bmatrix} \times
$$

$$
\times
\begin{bmatrix} u_z \\ \phi_x \\ \phi_y \end{bmatrix}
+
\begin{bmatrix} -q_z \\ m_x \\ m_y \end{bmatrix}
=
\begin{bmatrix} 0 \\ 0 \\ 0 \end{bmatrix},
\qquad (7.23)
$$

or symbolically as

$$
\mathcal{L}_1^{\mathrm{T}} D \mathcal{L}_1 u + b = 0. \qquad (7.24)
$$

Table 7.1 summarizes the different formulations of the basic equations for a thick plate.

The general formulations of the basic equations for a thick plate as given in Table 7.1 can be slightly modified to avoid some esthetic appeals and to have a similar structure as in the case of the Timoshenko beam, see Table 4.6. The kinematics equation remains unchanged while the minus signs in the constitutive equation (see the matrix of generalized strains) can be eliminated:

Table 7.1 Different formulations of the basic equations for a thick plate (bending perpendicular to the x-y plane). E: Young's modulus; ν: Poisson's ratio; G: shear modulus; q_z: area-specific distributed force; m: area-specific distributed moment; h plate thickness; k_s: shear correction factor; M^n: length-specific moment; Q^n: length-specific shear force; e: generalized strains; s: generalized stresses

Specific formulation	General formulation

Kinematics

$$\begin{bmatrix} \frac{\partial \phi_y}{\partial x} \\ -\frac{\partial \phi_x}{\partial y} \\ \frac{\partial \phi_y}{\partial y} - \frac{\partial \phi_x}{\partial x} \\ \phi_y + \frac{\partial u_z}{\partial x} \\ -\phi_x + \frac{\partial u_z}{\partial y} \end{bmatrix} = \begin{bmatrix} 0 & 0 & \frac{\partial}{\partial x} \\ 0 & -\frac{\partial}{\partial y} & 0 \\ 0 & -\frac{\partial}{\partial x} & \frac{\partial}{\partial y} \\ \frac{\partial}{\partial x} & 0 & 1 \\ \frac{\partial}{\partial y} & -1 & 0 \end{bmatrix} \begin{bmatrix} u_z \\ \phi_x \\ \phi_y \end{bmatrix}$$

$$e = \mathcal{L}_1 u$$

Constitution

$$\begin{bmatrix} M_x^n \\ M_y^n \\ M_{xy}^n \\ -Q_x^n \\ -Q_y^n \end{bmatrix} = \begin{bmatrix} \underbrace{\frac{Eh^3}{12(1-\nu^2)}\begin{bmatrix} 1 & \nu & 0 \\ \nu & 1 & 0 \\ 0 & 0 & \frac{1-\nu}{2} \end{bmatrix}}_{D_b} & \begin{bmatrix} 0 & 0 \\ 0 & 0 \\ 0 & 0 \end{bmatrix} \\ \begin{bmatrix} 0 & 0 & 0 \\ 0 & 0 & 0 \end{bmatrix} & \underbrace{-k_s Gh \begin{bmatrix} 1 & 0 \\ 0 & 1 \end{bmatrix}}_{D_s} \end{bmatrix} \begin{bmatrix} \frac{\partial \phi_y}{\partial x} \\ -\frac{\partial \phi_x}{\partial y} \\ \frac{\partial \phi_y}{\partial y} - \frac{\partial \phi_x}{\partial x} \\ \phi_y + \frac{\partial u_z}{\partial x} \\ -\phi_x + \frac{\partial u_z}{\partial y} \end{bmatrix}$$

$$s = De$$

Equilibrium

$$\begin{bmatrix} 0 & 0 & 0 & \frac{\partial}{\partial x} & \frac{\partial}{\partial y} \\ 0 & -\frac{\partial}{\partial y} & -\frac{\partial}{\partial x} & 0 & -1 \\ \frac{\partial}{\partial x} & 0 & \frac{\partial}{\partial y} & 1 & 0 \end{bmatrix} \begin{bmatrix} M_x^n \\ M_y^n \\ M_{xy}^n \\ -Q_x^n \\ -Q_y^n \end{bmatrix} + \begin{bmatrix} -q_z \\ m_x \\ m_y \end{bmatrix} = \begin{bmatrix} 0 \\ 0 \\ 0 \end{bmatrix}$$

$$\mathcal{L}_1^T s + b = 0$$

PDE

$$\begin{bmatrix} -D_s\left(\frac{\partial^2}{\partial x^2} + \frac{\partial^2}{\partial y^2}\right) & D_s\frac{\partial}{\partial y} & -D_s\frac{\partial}{\partial x} \\ D_s\frac{\partial}{\partial y} & D_b\left(\frac{1-\nu}{2}\frac{\partial^2}{\partial x^2} + \frac{\partial^2}{\partial y^2}\right) - D_s & -\frac{1+\nu}{2}D_b\frac{\partial^2}{\partial x \partial y} \\ -D_s\frac{\partial}{\partial x} & -\frac{1+\nu}{2}D_b\frac{\partial^2}{\partial x \partial y} & D_b\left(\frac{\partial^2}{\partial x^2} + \frac{1-\nu}{2}\frac{\partial^2}{\partial y^2}\right) - D_s \end{bmatrix}$$

$$\begin{bmatrix} u_z \\ \phi_x \\ \phi_y \end{bmatrix} + \begin{bmatrix} -q_z \\ m_x \\ m_y \end{bmatrix} = \begin{bmatrix} 0 \\ 0 \\ 0 \end{bmatrix}$$

$$\mathcal{L}_1^T D \mathcal{L}_1 u + b = 0$$

$$
\begin{bmatrix} 1 & 0 & 0 & 0 & 0 \\ 0 & 1 & 0 & 0 & 0 \\ 0 & 0 & 1 & 0 & 0 \\ 0 & 0 & 0 & -1 & 0 \\ 0 & 0 & 0 & 0 & -1 \end{bmatrix} \begin{bmatrix} M_x^n \\ M_y^n \\ M_{xy}^n \\ Q_x^n \\ Q_y^n \end{bmatrix} = \begin{bmatrix} 1 & 0 & 0 & 0 & 0 \\ 0 & 1 & 0 & 0 & 0 \\ 0 & 0 & 1 & 0 & 0 \\ 0 & 0 & 0 & -1 & 0 \\ 0 & 0 & 0 & 0 & -1 \end{bmatrix} \left[\begin{array}{c|c} D_b \begin{bmatrix} \cdots \end{bmatrix} & \begin{bmatrix} \cdots \end{bmatrix} \\ \hline \begin{bmatrix} \cdots \end{bmatrix} & D_s \begin{bmatrix} \cdots \end{bmatrix} \end{array} \right] \begin{bmatrix} \dfrac{\partial \phi_y}{\partial x} \\ -\dfrac{\partial \phi_x}{\partial y} \\ \dfrac{\partial \phi_y}{\partial y} - \dfrac{\partial \phi_x}{\partial x} \\ \gamma_{xz} \\ \gamma_{yz} \end{bmatrix}, \quad (7.25)
$$

The diagonal matrix $\lceil 1\ 1\ 1\ -1\ -1 \rfloor$ can be eliminated from the last equation to obtain the modified constitutive law in matrix notation:

$$
\begin{bmatrix} M_x^n \\ M_y^n \\ M_{xy}^n \\ Q_x^n \\ Q_y^n \end{bmatrix} = \left[\begin{array}{c|c} D_b \begin{bmatrix} 1 & \nu & 0 \\ \nu & 1 & 0 \\ 0 & 0 & \frac{1-\nu}{2} \end{bmatrix} & \begin{bmatrix} 0 & 0 \\ 0 & 0 \\ 0 & 0 \end{bmatrix} \\ \hline \begin{bmatrix} 0 & 0 & 0 \\ 0 & 0 & 0 \end{bmatrix} & D_s \begin{bmatrix} 1 & 0 \\ 0 & 1 \end{bmatrix} \end{array} \right] \begin{bmatrix} \dfrac{\partial \phi_y}{\partial x} \\ -\dfrac{\partial \phi_x}{\partial y} \\ \dfrac{\partial \phi_y}{\partial y} - \dfrac{\partial \phi_x}{\partial x} \\ \gamma_{xz} \\ \gamma_{yz} \end{bmatrix}. \quad (7.26)
$$

The next step is to have a closer look on the equilibrium equation, i.e.,

$$
\begin{bmatrix} 0 & 0 & 0 & \dfrac{\partial}{\partial x} & \dfrac{\partial}{\partial y} \\ 0 & -\dfrac{\partial}{\partial y} & -\dfrac{\partial}{\partial x} & 0 & -1 \\ \dfrac{\partial}{\partial x} & 0 & \dfrac{\partial}{\partial y} & 1 & 0 \end{bmatrix} \begin{bmatrix} M_x^n \\ M_y^n \\ M_{xy}^n \\ -Q_x^n \\ -Q_y^n \end{bmatrix} + \begin{bmatrix} -q_z \\ m_x \\ m_y \end{bmatrix} = \begin{bmatrix} 0 \\ 0 \\ 0 \end{bmatrix}, \quad (7.27)
$$

or again re-written based on the diagonal matrices to extract the minus signs:

$$
\begin{bmatrix} 0 & 0 & 0 & \dfrac{\partial}{\partial x} & \dfrac{\partial}{\partial y} \\ 0 & -\dfrac{\partial}{\partial y} & -\dfrac{\partial}{\partial x} & 0 & -1 \\ \dfrac{\partial}{\partial x} & 0 & \dfrac{\partial}{\partial y} & 1 & 0 \end{bmatrix} \begin{bmatrix} 1 & 0 & 0 & 0 & 0 \\ 0 & 1 & 0 & 0 & 0 \\ 0 & 0 & 1 & 0 & 0 \\ 0 & 0 & 0 & -1 & 0 \\ 0 & 0 & 0 & 0 & -1 \end{bmatrix} \begin{bmatrix} M_x^n \\ M_y^n \\ M_{xy}^n \\ Q_x^n \\ Q_y^n \end{bmatrix} + \begin{bmatrix} -1 & 0 & 0 \\ 0 & 1 & 0 \\ 0 & 0 & 1 \end{bmatrix} \begin{bmatrix} q_z \\ m_x \\ m_y \end{bmatrix} = \begin{bmatrix} 0 \\ 0 \\ 0 \end{bmatrix}. \quad (7.28)
$$

Let us now multiply the first two matrices and then multiply the resulting equation with the (3×3) diagonal matrix from the left-hand side:

$$
\begin{bmatrix} -1 & 0 & 0 \\ 0 & 1 & 0 \\ 0 & 0 & 1 \end{bmatrix}
\begin{bmatrix} 0 & 0 & 0 & -\dfrac{\partial}{\partial x} & -\dfrac{\partial}{\partial y} \\[2mm] 0 & -\dfrac{\partial}{\partial y} & -\dfrac{\partial}{\partial x} & 0 & 1 \\[2mm] \dfrac{\partial}{\partial x} & 0 & \dfrac{\partial}{\partial y} & -1 & 0 \end{bmatrix}
\begin{bmatrix} 1 & 0 & 0 & 0 & 0 \\ 0 & 1 & 0 & 0 & 0 \\ 0 & 0 & 1 & 0 & 0 \\ 0 & 0 & 0 & -1 & 0 \\ 0 & 0 & 0 & 0 & -1 \end{bmatrix}
\begin{bmatrix} M_x^n \\ M_y^n \\ M_{xy}^n \\ Q_x^n \\ Q_y^n \end{bmatrix} +
$$

$$
\begin{bmatrix} -1 & 0 & 0 \\ 0 & 1 & 0 \\ 0 & 0 & 1 \end{bmatrix}
\begin{bmatrix} -1 & 0 & 0 \\ 0 & 1 & 0 \\ 0 & 0 & 1 \end{bmatrix}
\begin{bmatrix} q_z \\ m_x \\ m_y \end{bmatrix} =
\begin{bmatrix} 0 \\ 0 \\ 0 \end{bmatrix}. \tag{7.29}
$$

Or finally as the modified expression of the equilibrium equation:

$$
\begin{bmatrix} 0 & 0 & 0 & \dfrac{\partial}{\partial x} & \dfrac{\partial}{\partial y} \\[2mm] 0 & -\dfrac{\partial}{\partial y} & -\dfrac{\partial}{\partial x} & 0 & 1 \\[2mm] \dfrac{\partial}{\partial x} & 0 & \dfrac{\partial}{\partial y} & -1 & 0 \end{bmatrix}
\begin{bmatrix} M_x^n \\ M_y^n \\ M_{xy}^n \\ Q_x^n \\ Q_y^n \end{bmatrix} +
\begin{bmatrix} q_z \\ m_x \\ m_y \end{bmatrix} =
\begin{bmatrix} 0 \\ 0 \\ 0 \end{bmatrix}. \tag{7.30}
$$

Combining the three basic equations results again in the system of partial differential equations:

$$
\begin{bmatrix} D_s\left(\dfrac{\partial^2}{\partial x^2}+\dfrac{\partial^2}{\partial y^2}\right) & -D_s\dfrac{\partial}{\partial y} & D_s\dfrac{\partial}{\partial x} \\[3mm] D_s\dfrac{\partial}{\partial y} & D_b\left(\dfrac{1-\nu}{2}\dfrac{\partial^2}{\partial x^2}+\dfrac{\partial^2}{\partial y^2}\right)-D_s & -\dfrac{1+\nu}{2}D_b\dfrac{\partial^2}{\partial x\partial y} \\[3mm] -D_s\dfrac{\partial}{\partial x} & -\dfrac{1+\nu}{2}D_b\dfrac{\partial^2}{\partial x\partial y} & D_b\left(\dfrac{\partial^2}{\partial x^2}+\dfrac{1-\nu}{2}\dfrac{\partial^2}{\partial y^2}\right)-D_s \end{bmatrix} \times
$$

$$
\times \begin{bmatrix} u_z \\ \phi_x \\ \phi_y \end{bmatrix} + \begin{bmatrix} q_z \\ m_x \\ m_y \end{bmatrix} = \begin{bmatrix} 0 \\ 0 \\ 0 \end{bmatrix}. \tag{7.31}
$$

The modified basic equations, i.e., 'without the minus signs', are summarized in Table 7.2.

Table 7.2 Alternative formulations of the basic equations for a thick plate

Specific formulation	General formulation

Kinematics

$$\begin{bmatrix} \dfrac{\partial \phi_y}{\partial x} \\[2mm] -\dfrac{\partial \phi_x}{\partial y} \\[2mm] \dfrac{\partial \phi_y}{\partial y} - \dfrac{\partial \phi_x}{\partial x} \\[2mm] \phi_y + \dfrac{\partial u_z}{\partial x} \\[2mm] -\phi_x + \dfrac{\partial u_z}{\partial y} \end{bmatrix} = \begin{bmatrix} 0 & 0 & \dfrac{\partial}{\partial x} \\[2mm] 0 & -\dfrac{\partial}{\partial y} & 0 \\[2mm] 0 & -\dfrac{\partial}{\partial x} & \dfrac{\partial}{\partial y} \\[2mm] \dfrac{\partial}{\partial x} & 0 & 1 \\[2mm] \dfrac{\partial}{\partial y} & -1 & 0 \end{bmatrix} \begin{bmatrix} u_z \\ \phi_x \\ \phi_y \end{bmatrix} \qquad\qquad e = \mathcal{L}_1 u$$

Constitution

$$\begin{bmatrix} M_x^n \\ M_y^n \\ M_{xy}^n \\ Q_x^n \\ Q_y^n \end{bmatrix} = \left[\begin{array}{c|c} \underbrace{\dfrac{Eh^3}{12(1-\nu^2)}}_{D_b}\begin{bmatrix} 1 & \nu & 0 \\ \nu & 1 & 0 \\ 0 & 0 & \frac{1-\nu}{2} \end{bmatrix} & \begin{bmatrix} 0 & 0 \\ 0 & 0 \\ 0 & 0 \end{bmatrix} \\ \hline \begin{bmatrix} 0 & 0 & 0 \\ 0 & 0 & 0 \end{bmatrix} & \underbrace{k_s G h}_{D_s}\begin{bmatrix} 1 & 0 \\ 0 & 1 \end{bmatrix} \end{array}\right] \begin{bmatrix} \frac{\partial \phi_y}{\partial x} \\ -\frac{\partial \phi_x}{\partial y} \\ \frac{\partial \phi_y}{\partial y} - \frac{\partial \phi_x}{\partial x} \\ \phi_y + \frac{\partial u_z}{\partial x} \\ -\phi_x + \frac{\partial u_z}{\partial y} \end{bmatrix} \qquad s^* = D^* e$$

Equilibrium

$$\begin{bmatrix} 0 & 0 & 0 & \frac{\partial}{\partial x} & \frac{\partial}{\partial y} \\ 0 & -\frac{\partial}{\partial y} & -\frac{\partial}{\partial x} & 0 & 1 \\ \frac{\partial}{\partial x} & 0 & \frac{\partial}{\partial y} & -1 & 0 \end{bmatrix} \begin{bmatrix} M_x^n \\ M_y^n \\ M_{xy}^n \\ Q_x^n \\ Q_y^n \end{bmatrix} + \begin{bmatrix} q_z \\ m_x \\ m_y \end{bmatrix} = \begin{bmatrix} 0 \\ 0 \\ 0 \end{bmatrix} \qquad \mathcal{L}_{1*}^T s^* + b^* = 0$$

PDE

$$\begin{bmatrix} D_s\left(\dfrac{\partial^2}{\partial x^2} + \dfrac{\partial^2}{\partial y^2}\right) & -D_s\dfrac{\partial}{\partial y} & D_s\dfrac{\partial}{\partial x} \\[2mm] D_s\dfrac{\partial}{\partial y} & D_b\left(\dfrac{1-\nu}{2}\dfrac{\partial^2}{\partial x^2} + \dfrac{\partial^2}{\partial y^2}\right) - D_s & -\dfrac{1+\nu}{2}D_b\dfrac{\partial^2}{\partial x \partial y} \\[2mm] -D_s\dfrac{\partial}{\partial x} & -\dfrac{1+\nu}{2}D_b\dfrac{\partial^2}{\partial x \partial y} & D_b\left(\dfrac{\partial^2}{\partial x^2} + \dfrac{1-\nu}{2}\dfrac{\partial^2}{\partial y^2}\right) - D_s \end{bmatrix} \begin{bmatrix} u_z \\ \phi_x \\ \phi_y \end{bmatrix} + \begin{bmatrix} q_z \\ m_x \\ m_y \end{bmatrix} = \begin{bmatrix} 0 \\ 0 \\ 0 \end{bmatrix} \qquad \mathcal{L}_{1*}^T D^* \mathcal{L}_1 u + b^* = 0$$

References

1. Blaauwendraad J (2010) Plates and FEM: surprises and pitfalls. Springer, Dordrecht
2. Mindlin RD (1951) Influence of rotary inertia and shear on flexural motions isotropic, elastic plates. J Appl Mech-T ASME 18:1031–1036
3. Reddy JN (2006) An introduction to the finite element method. McGraw Hill, Singapore
4. Reissner E (1945) The effect of transverse shear deformation on the bending of elastic plates. J Appl Mech-T ASME 12:A68–A77
5. Ventsel E, Krauthammer T (2001) Thin plates and shells: theory, analysis, and applications. Marcel Dekker, New York
6. Wang CM, Reddy JN, Lee KH (2000) Shear deformable beams and plates: relationships with classical solution. Elsevier, Oxford

Chapter 8
Three-Dimensional Solids

Abstract This chapter covers the continuum mechanical description of solid or three-dimensional members. Based on the three basic equations of continuum mechanics, i.e., the kinematics relationship, the constitutive law, and the equilibrium equation, the partial differential equation, which describes the physical problem, is derived.

8.1 Introduction

A solid element is defined as a three-dimensional member, as schematically shown in Fig. 8.1, where all dimensions have a similar magnitude. It can be seen as a three-dimensional extension or generalization of the plane elasticity element. The following derivations are restricted to some simplifications:

- the material is isotropic, homogenous and linear-elastic according to Hooke's law for a three-dimensional stress and strain state,
- only members with 6 faces, 12 edges, 8 vertices (hexahedra) are considered.

The analogies between the rod, plane elasticity theories and three-dimensional elements are summarized in Table 8.1.

8.2 Kinematics

The kinematics or strain-displacement relations extract the strain field contained in a displacement field. Using engineering definitions of strain, the following relations can be obtained [2, 3]:

© The Author(s), under exclusive license to Springer Nature Switzerland AG 2020   81
A. Öchsner, *Partial Differential Equations of Classical Structural Members*,
SpringerBriefs in Continuum Mechanics,
https://doi.org/10.1007/978-3-030-35311-7_8

Fig. 8.1 General
configuration for a
three-dimensional problem

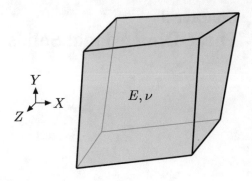

Table 8.1 Difference between rod, plane and three-dimensional element

Rod	Plane element	3D element
1D	2D	3D
Deformation along principal axis	In-plane deformation	Spatial deformation
u_x	u_x, u_y	u_x, u_y, u_z

$$\varepsilon_x = \frac{\partial u_x}{\partial x} \; ; \; \varepsilon_y = \frac{\partial u_y}{\partial y} \; ; \; \varepsilon_z = \frac{\partial u_z}{\partial z} \; ; \; \varepsilon_{xy} = \frac{1}{2}\left(\frac{\partial u_x}{\partial y} + \frac{\partial u_y}{\partial x}\right) \; ; \quad (8.1)$$

$$\varepsilon_{xz} = \frac{1}{2}\left(\frac{\partial u_x}{\partial z} + \frac{\partial u_z}{\partial x}\right) \; ; \; \varepsilon_{yz} = \frac{1}{2}\left(\frac{\partial u_y}{\partial z} + \frac{\partial u_z}{\partial y}\right) . \quad (8.2)$$

In matrix notation, these six relationships can be written as

$$\begin{bmatrix} \varepsilon_x \\ \varepsilon_y \\ \varepsilon_z \\ 2\varepsilon_{xy} \\ 2\varepsilon_{yz} \\ 2\varepsilon_{xz} \end{bmatrix} = \begin{bmatrix} \frac{\partial}{\partial x} & 0 & 0 \\ 0 & \frac{\partial}{\partial y} & 0 \\ 0 & 0 & \frac{\partial}{\partial z} \\ \frac{\partial}{\partial y} & \frac{\partial}{\partial x} & 0 \\ 0 & \frac{\partial}{\partial z} & \frac{\partial}{\partial y} \\ \frac{\partial}{\partial z} & 0 & \frac{\partial}{\partial x} \end{bmatrix} \begin{bmatrix} u_x \\ u_y \\ u_z \end{bmatrix} , \quad (8.3)$$

or symbolically as

$$\varepsilon = \mathcal{L}_1 u . \quad (8.4)$$

8.3 Constitution

The generalized Hooke's law for a linear-elastic isotropic material based on the Young's modulus E and Poisson's ratio ν can be written for a constant temperature with all components as

$$
\begin{bmatrix} \sigma_x \\ \sigma_y \\ \sigma_z \\ \sigma_{xy} \\ \sigma_{yz} \\ \sigma_{xz} \end{bmatrix} = \frac{E}{(1+\nu)(1-2\nu)} \begin{bmatrix} 1-\nu & \nu & \nu & 0 & 0 & 0 \\ \nu & 1-\nu & \nu & 0 & 0 & 0 \\ \nu & \nu & 1-\nu & 0 & 0 & 0 \\ 0 & 0 & 0 & \frac{1-2\nu}{2} & 0 & 0 \\ 0 & 0 & 0 & 0 & \frac{1-2\nu}{2} & 0 \\ 0 & 0 & 0 & 0 & 0 & \frac{1-2\nu}{2} \end{bmatrix} \begin{bmatrix} \varepsilon_x \\ \varepsilon_y \\ \varepsilon_z \\ 2\varepsilon_{xy} \\ 2\varepsilon_{yz} \\ 2\varepsilon_{xz} \end{bmatrix} ,
$$

(8.5)

or in matrix notation as

$$
\sigma = C\epsilon ,
$$

(8.6)

where C is the so-called elasticity matrix. It should be noted here that the engineering shear strain $\gamma_{ij} = 2\varepsilon_{ij}$ (for $i \neq j$) is used in the formulation of Eq. (8.5), see Sect. 4.1 for further details. Rearranging the elastic stiffness form given in Eq. (8.5) for the strains gives the elastic compliance form

$$
\begin{bmatrix} \varepsilon_x \\ \varepsilon_y \\ \varepsilon_z \\ 2\varepsilon_{xy} \\ 2\varepsilon_{yz} \\ 2\varepsilon_{xz} \end{bmatrix} = \frac{1}{E} \begin{bmatrix} 1 & -\nu & -\nu & 0 & 0 & 0 \\ -\nu & 1 & -\nu & 0 & 0 & 0 \\ -\nu & -\nu & 1 & 0 & 0 & 0 \\ 0 & 0 & 0 & 2(1+\nu) & 0 & 0 \\ 0 & 0 & 0 & 0 & 2(1+\nu) & 0 \\ 0 & 0 & 0 & 0 & 0 & 2(1+\nu) \end{bmatrix} \begin{bmatrix} \sigma_x \\ \sigma_y \\ \sigma_z \\ \sigma_{xy} \\ \sigma_{yz} \\ \sigma_{xz} \end{bmatrix} ,
$$

(8.7)

or in matrix notation as

$$
\epsilon = D\sigma ,
$$

(8.8)

where $D = C^{-1}$ is the so-called elastic compliance matrix. The general characteristic of Hooke's law in the form of Eqs. (8.6) and (8.8) is that two independent material parameters are used. In addition to Young's modulus E and Poisson's ratio ν, other elastic parameters can be used to form the set of two independent material parameters and the following Table 8.2 summarizes the conversion between the common material parameters.

Table 8.2 Conversion of elastic constants: λ, μ: Lamé's constants; K: bulk modulus; G: shear modulus; E: Young's modulus; ν: Poisson's ratio, [1]

	λ, μ	E, ν	μ, ν	E, μ	K, ν	G, ν	K, G
λ	λ	$\frac{\nu E}{(1+\nu)(1-2\nu)}$	$\frac{2\mu\nu}{1-2\nu}$	$\frac{\mu(E-2\mu)}{3\mu-E}$	$\frac{3K\nu}{1+\nu}$	$\frac{2G\nu}{1-2\nu}$	$K - \frac{2G}{3}$
μ	μ	$\frac{E}{2(1+\nu)}$	μ	μ	$\frac{3K(1-2\nu)}{2(1+\nu)}$	μ	μ
K	$\lambda + \frac{2}{3}\mu$	$\frac{E}{3(1-2\nu)}$	$\frac{2\mu(1+\nu)}{3(1-2\nu)}$	$\frac{\mu E}{3(3\mu-E)}$	K	$\frac{2G(1+\nu)}{3(1-2\nu)}$	K
E	$\frac{\mu(3\lambda+2\mu)}{\lambda+\mu}$	E	$2\mu(1+\nu)$	E	$3K(1-2\nu)$	$2G(1+\nu)$	$\frac{9KG}{3K+G}$
ν	$\frac{\lambda}{2(\lambda+\mu)}$	ν	ν	$\frac{E}{2\mu} - 1$	ν	ν	$\frac{3K-2G}{2(3K+G)}$
G	μ	$\frac{E}{2(1+\nu)}$	μ	G	$\frac{3K(1-2\nu)}{2(1+\nu)}$	G	G

8.4 Equilibrium

Figure. 8.2 shows the normal and shear stresses which are acting on a differential volume element in the x-direction. All forces are drawn in their positive direction at each cut face. A positive cut face is obtained if the outward surface normal is directed in the positive direction of the corresponding coordinate axis. This means that the right-hand face in Fig. 8.2 is positive and the force $(\sigma_x + \frac{\partial \sigma_x}{\partial x}dx)dydz$ is oriented in the positive x-direction. In a similar way, the top face is positive, i.e. the outward surface normal is directed in the positive y-direction, and the shear force[1] is oriented in the positive x-direction. Since the volume element is assumed to be in equilibrium, forces resulting from stresses on the sides of the cuboid and from the body forces f_i ($i = x, y, z$) must be balanced. These body forces are defined as forces per unit volume which can be produced by gravity,[2] acceleration, magnetic fields, and so on.

The static equilibrium of forces in the x-direction based on the seven force components—two normal forces, four shear forces and one body force—indicated in Fig. 8.2 gives after canceling with $dV = dxdydz$:

$$\frac{\partial \sigma_x}{\partial x} + \frac{\partial \sigma_{yx}}{\partial y} + \frac{\partial \sigma_{zx}}{\partial z} + f_x = 0. \qquad (8.9)$$

Based on the same approach, similar equations can be specified in the y- and z-direction:

---

[1]In the case of a shear force σ_{ij}, the first index i indicates that the stress acts on a plane normal to the i-axis and the second index j denotes the direction in which the stress acts.

[2]If gravity is acting, the body force f results as the product of density times standard gravity: $f = \frac{F}{V} = \frac{mg}{V} = \frac{m}{V}g = \varrho g$. The units can be checked by consideration of $1\,\text{N} = 1\,\frac{\text{mkg}}{\text{s}^2}$.

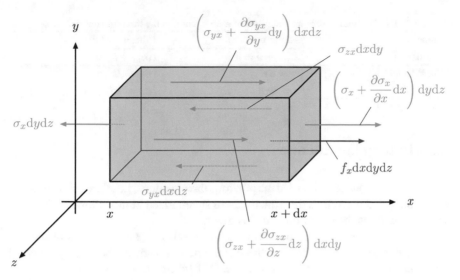

Fig. 8.2 Stress and body forces which act on a differential volume element in x-direction

$$\frac{\partial \sigma_y}{\partial y} + \frac{\partial \sigma_{yx}}{\partial x} + \frac{\partial \sigma_{yz}}{\partial z} + f_y = 0, \tag{8.10}$$

$$\frac{\partial \sigma_z}{\partial z} + \frac{\partial \sigma_{xz}}{\partial x} + \frac{\partial \sigma_{yz}}{\partial y} + f_z = 0. \tag{8.11}$$

These three balance equations can be written in matrix notation as

$$\begin{bmatrix} \dfrac{\partial}{\partial x} & 0 & 0 & \dfrac{\partial}{\partial y} & 0 & \dfrac{\partial}{\partial z} \\[2mm] 0 & \dfrac{\partial}{\partial y} & 0 & \dfrac{\partial}{\partial x} & \dfrac{\partial}{\partial z} & 0 \\[2mm] 0 & 0 & \dfrac{\partial}{\partial z} & 0 & \dfrac{\partial}{\partial y} & \dfrac{\partial}{\partial x} \end{bmatrix} \begin{bmatrix} \sigma_x \\ \sigma_y \\ \sigma_z \\ \sigma_{xy} \\ \sigma_{yz} \\ \sigma_{xz} \end{bmatrix} + \begin{bmatrix} f_x \\ f_y \\ f_z \end{bmatrix} = \begin{bmatrix} 0 \\ 0 \\ 0 \end{bmatrix}, \tag{8.12}$$

or in symbolic notation:

$$\mathcal{L}_1^{\mathrm{T}} \boldsymbol{\sigma} + \boldsymbol{b} = \mathbf{0}, \tag{8.13}$$

where $\mathcal{L}_1$ is the differential operator matrix and $\boldsymbol{b}$ the column matrix of body forces.

Table 8.3 Fundamental governing equations of a continuum in the three-dimensional case

Name	Matrix notation	Tensor notation
Equilibrium	$\mathcal{L}_1^{\mathrm{T}} \sigma + b = 0$	$\sigma_{ij,i} + b_j = 0$
Constitution	$\sigma = C \epsilon$	$\sigma_{ij} = C_{ijkl} \varepsilon_{kl}$
Kinematics	$\varepsilon = \mathcal{L}_1 u$	$\varepsilon_{ij} = \frac{1}{2} \left(u_{i,j} + u_{j,i} \right)$

8.5   Differential Equation

The basic equations introduced in the previous three sections, i.e., the equilibrium, the constitutive and the kinematics equation, are summarized in the following Table 8.3 where in addition the tensor notation[3] is given.

For the solution of the 15 unknown spatial functions (3 components of the displacement vector, 6 components of the symmetric strain tensor and 6 components of the symmetric stress tensor), a set of 15 scalar field equations is available:

- Equilibrium: 3,
- Constitution: 6,
- Kinematics: 6.

Furthermore, the boundary conditions are given:

$$u \quad \text{on} \quad \Gamma_u \,, \tag{8.14}$$

$$t \quad \text{on} \quad \Gamma_t \,, \tag{8.15}$$

where Γ_u is the part of the boundary where a displacement boundary condition is prescribed and Γ_t is the part of the boundary where a traction boundary condition, i.e. external force per unit area, is prescribed with $t_j = \sigma_{ij} n_j$, where n_j are the components of the normal vector.

The 15 scalar field equations can be combined to eliminate the stress and strain fields. As a result, three scalar field equations for the three scalar displacement fields are obtained. These equations are called the Lamé–Navier equations and can be derived as follows:

Introducing the constitutive equation according to (8.6) in the equilibrium equation (8.13) gives:

$$\mathcal{L}_1^{\mathrm{T}} C \varepsilon + b = 0 \,. \tag{8.16}$$

Introducing the kinematics relations in the last equation according to (8.4) finally gives the Lamé–Navier equations:

---

[3] A differentiation is there indicated by the use of a comma: The first index refers to the component and the comma indicates the partial derivative with respect to the second subscript corresponding to the relevant coordinate axis, [2].

Table 8.4 Different formulations of the basic equations for three-dimensional elasticity. E: Young's modulus; ν: Poisson's ratio; f: volume-specific force [4]

Specific formulation	General formulation

Kinematics

$$\begin{bmatrix} \varepsilon_x \\ \varepsilon_y \\ \varepsilon_z \\ 2\varepsilon_{xy} \\ 2\varepsilon_{yz} \\ 2\varepsilon_{xz} \end{bmatrix} = \begin{bmatrix} \frac{\partial}{\partial x} & 0 & 0 \\ 0 & \frac{\partial}{\partial y} & 0 \\ 0 & 0 & \frac{\partial}{\partial z} \\ \frac{\partial}{\partial y} & \frac{\partial}{\partial x} & 0 \\ 0 & \frac{\partial}{\partial z} & \frac{\partial}{\partial y} \\ \frac{\partial}{\partial z} & 0 & \frac{\partial}{\partial x} \end{bmatrix} \begin{bmatrix} u_x \\ u_y \\ u_z \end{bmatrix}$$

$\varepsilon = \mathcal{L}_1 u$

Constitution

$$\begin{bmatrix} \sigma_x \\ \sigma_y \\ \sigma_z \\ \sigma_{xy} \\ \sigma_{yz} \\ \sigma_{xz} \end{bmatrix} = \frac{E}{(1+\nu)(1-2\nu)} \begin{bmatrix} 1-\nu & \nu & \nu & 0 & 0 & 0 \\ \nu & 1-\nu & \nu & 0 & 0 & 0 \\ \nu & \nu & 1-\nu & 0 & 0 & 0 \\ 0 & 0 & 0 & \frac{1-2\nu}{2} & 0 & 0 \\ 0 & 0 & 0 & 0 & \frac{1-2\nu}{2} & 0 \\ 0 & 0 & 0 & 0 & 0 & \frac{1-2\nu}{2} \end{bmatrix} \begin{bmatrix} \varepsilon_x \\ \varepsilon_y \\ \varepsilon_z \\ 2\varepsilon_{xy} \\ 2\varepsilon_{yz} \\ 2\varepsilon_{xz} \end{bmatrix}$$

$\sigma = C\varepsilon$

Equilibrium

$$\begin{bmatrix} \frac{\partial}{\partial x} & 0 & 0 & \frac{\partial}{\partial y} & 0 & \frac{\partial}{\partial z} \\ 0 & \frac{\partial}{\partial y} & 0 & \frac{\partial}{\partial x} & \frac{\partial}{\partial z} & 0 \\ 0 & 0 & \frac{\partial}{\partial z} & 0 & \frac{\partial}{\partial y} & \frac{\partial}{\partial x} \end{bmatrix} \begin{bmatrix} \sigma_x \\ \sigma_y \\ \sigma_z \\ \sigma_{xy} \\ \sigma_{yz} \\ \sigma_{xz} \end{bmatrix} + \begin{bmatrix} f_x \\ f_y \\ f_z \end{bmatrix} = \begin{bmatrix} 0 \\ 0 \\ 0 \end{bmatrix}$$

$\mathcal{L}_1^T \sigma + b = 0$

PDE

$$\frac{E}{(1+\nu)(1-2\nu)} \begin{bmatrix} \cdots & \cdots & \cdots \\ \cdots & \cdots & \cdots \\ \cdots & \cdots & \cdots \end{bmatrix} \begin{bmatrix} u_x \\ u_y \\ u_z \end{bmatrix} + \begin{bmatrix} f_x \\ f_y \\ f_z \end{bmatrix} = \begin{bmatrix} 0 \\ 0 \\ 0 \end{bmatrix}$$

$\mathcal{L}_1^T C \mathcal{L}_1 u + b = 0$

with $\begin{bmatrix} \cdots & \cdots & \cdots \\ \cdots & \cdots & \cdots \\ \cdots & \cdots & \cdots \end{bmatrix} =$

$$\begin{bmatrix} (1-\nu)\frac{d^2}{dx^2} + \left(\frac{1}{2}-\nu\right)\left(\frac{d^2}{dy^2}+\frac{d^2}{dz^2}\right) & \nu\frac{d^2}{dxdy} + \left(\frac{1}{2}-\nu\right)\frac{d^2}{dxdy} & \nu\frac{d^2}{dxdz} + \left(\frac{1}{2}-\nu\right)\frac{d^2}{dxdz} \\ \nu\frac{d^2}{dxdy} + \left(\frac{1}{2}-\nu\right)\frac{d^2}{dxdy} & (1-\nu)\frac{d^2}{dy^2} + \left(\frac{1}{2}-\nu\right)\left(\frac{d^2}{dx^2}+\frac{d^2}{dz^2}\right) & \nu\frac{d^2}{dydz} + \left(\frac{1}{2}-\nu\right)\frac{d^2}{dydz} \\ \nu\frac{d^2}{dxdz} + \left(\frac{1}{2}-\nu\right)\frac{d^2}{dxdz} & \nu\frac{d^2}{dydz} + \left(\frac{1}{2}-\nu\right)\frac{d^2}{dydz} & (1-\nu)\frac{d^2}{dz^2} + \left(\frac{1}{2}-\nu\right)\left(\frac{d^2}{dy^2}+\frac{d^2}{dx^2}\right) \end{bmatrix}$$

Table 8.5 Comparison of basic equations for rod, plane elasticity and three-dimensional elements

Rod	Plane elasticity	Three-Dimensional
Kinematics		
$\varepsilon_x(x) = \mathcal{L}_1\left(u_x(x)\right)$	$\varepsilon = \mathcal{L}_1 u$	$\varepsilon = \mathcal{L}_1 u$
Constitution		
$\sigma_x(x) = C\varepsilon_x(x)$	$\sigma = C\varepsilon$	$\sigma = C\varepsilon$
Equilibrium		
$\mathcal{L}_1\left(\sigma_x(x)\right) + b = 0$	$\mathcal{L}_1^{\mathrm{T}}\sigma + b = 0$	$\mathcal{L}_1^{\mathrm{T}}\sigma + b = 0$
PDE		
$\mathcal{L}_1\left(C\mathcal{L}_1\left(u_x(x)\right)\right) + b = 0$	$\mathcal{L}_1^{\mathrm{T}}C\mathcal{L}_1 u + b = 0$	$\mathcal{L}_1^{\mathrm{T}}C\mathcal{L}_1 u + b = 0$

$$\mathcal{L}_1^{\mathrm{T}}C\mathcal{L}_1 u + b = 0 \,. \tag{8.17}$$

Alternatively, the displacements may be substituted and the differential equations are obtained in terms of stresses. This formulation is known as the Beltrami–Michell equations. If the body forces vanish ($b = 0$), the partial differential equations in terms of stresses are called the Beltrami equations.

Table 8.4 summarizes different formulations of the basic equations for three-dimensional elasticity, once in their specific form and once in symbolic notation.

Table 8.5 shows a comparison between the basic equations for a rod, plane and 3D elasticity. It can be seen that the use of the differential operator $\mathcal{L}_1\{\dots\}$ allows to depict a simple analogy between these sets of equations.

References

1. Chen WF, Saleeb AF (1982) Constitutive equations for engineering materials. Volume 1: Elasticity and modelling. Wiley, New York
2. Chen WF, Han DJ (1988) Plasticity for structural engineers. Springer, New York
3. Eschenauer H, Olhoff N, Schnell W (1997) Applied structural mechanics: Fundamentals of elasticity, load-bearing structures, structural optimization. Springer, Berlin
4. Öchsner A (2014) Elasto-plasticity of frame structure elements: modeling and simulation of rods and beams. Springer, Berlin

Chapter 9
Introduction to Transient Problems: Rods or Bars

Abstract This chapter introduces to transient problems, i.e. problems where the state variables are time-dependent. The general treatment of transient problems is illustrated at the example of the rod or bar member.

9.1 Introduction

In extension to the explanations in Chap. 2, the *mass* of the rod is now considered and represented by its mass density ϱ. In addition, the distributed load $p_x(x, t)$ and the point load $F_x(t)$ are now functions of the *time t*, see Fig. 9.1.

9.2 Kinematics

The derivation of the kinematics relation is analogous to the approach presented in Sect. 2.2. Consideration in addition the time t gives:

$$\varepsilon_x(x, t) = \frac{\mathrm{d}u_x(x, t)}{\mathrm{d}x}. \tag{9.1}$$

9.3 Constitution

The constitutive description is based on Hooke's law as presented in Sect. 2.3. Inclusion of the time gives for the relation between stress and strain [2]:

$$\sigma_x(x, t) = E\varepsilon_x(x, t). \tag{9.2}$$

A. Öchsner, *Partial Differential Equations of Classical Structural Members*,
SpringerBriefs in Continuum Mechanics,
https://doi.org/10.1007/978-3-030-35311-7_9

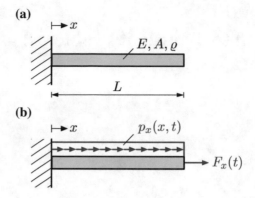

Fig. 9.1 General configuration of an axially loaded rod under consideration of time effects: **a** geometry and material property; **b** prescribed loads

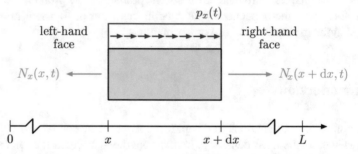

Fig. 9.2 Differential element of a rod under consideration of time effects with internal reactions and constant external distributed load

9.4   Equilibrium

The relationship between the external forces and internal reactions, as presented in Sect. 2.4, must be extended under consideration of the acceleration. Consider a differential element of length dx where the distributed load $p_x(t)$ and the cross-sectional area A are constant along the x-axis, see Fig. 9.2.

Application of Newton's second law in the x-direction gives:

$$- N_x(x, t) + p_x(t)dx + N_x(x + dx, t) = m \times a_x(x, t), \qquad (9.3)$$

where the acceleration can be expressed as $a_x(x, t) = \frac{d^2 u_x(x,t)}{dt^2}$. A first-order Taylor's series expansion of $N_x(x + dx)$ around point x, i.e.

$$N_x(x + dx) \approx N_x(x) + \frac{dN_x}{dx}\bigg|_x dx, \qquad (9.4)$$

Table 9.1 Fundamental governing equations of a rod for transient deformation along the x-axis

Expression	Equation
Kinematics	$\varepsilon_x(x,t) = \dfrac{\mathrm{d}u_x(x,t)}{\mathrm{d}x}$
Constitution	$\sigma_x(x,t) = E\varepsilon_x(x,t)$
Equilibrium	$\dfrac{\mathrm{d}N_x(x,t)}{\mathrm{d}x} = -p_x(x,t) + \dfrac{\gamma A}{g} \times \dfrac{\mathrm{d}^2 u_x(x,t)}{\mathrm{d}t^2}$

gives finally:

$$\frac{\mathrm{d}N_x(x,t)}{\mathrm{d}x} = -p_x(x,t) + \frac{m}{\mathrm{d}x} \times \frac{\mathrm{d}^2 u_x(x,t)}{\mathrm{d}t^2}, \tag{9.5}$$

with $m = \varrho \times V = \varrho \times A \times \mathrm{d}x = \frac{\gamma}{g} \times A \times \mathrm{d}x$. The mass and density related quantities are as follows[1]:

- ϱ: mass density (mass per unit volume) in $\frac{\mathrm{kg}}{\mathrm{m}^3}$,
- γ: weight density (weight per unit volume) in $\frac{\mathrm{N}}{\mathrm{m}^3}$,
- g: standard gravity or standard acceleration in $\frac{\mathrm{m}}{\mathrm{s}^2}$.

The three fundamental equations to describe the behavior of a rod element are summarized in Table 9.1.

9.5 Differential Equation

To derive the governing partial differential equation, the three fundamental equations given in Table 9.1 must be combined. Introducing the kinematics relation (9.1) into Hooke's law (9.2) gives:

$$\sigma_x(x,t) = E\frac{\mathrm{d}u_x(x,t)}{\mathrm{d}x}. \tag{9.6}$$

Considering in the last equation that a normal stress is defined as an acting force N_x over a cross-sectional area A:

$$\frac{N_x(x,t)}{A} = E\frac{\mathrm{d}u_x(x,t)}{\mathrm{d}x}. \tag{9.7}$$

The last equation can be differentiated with respect to the x-coordinate to give:

---

[1]Consider: $1\,\mathrm{N} = 1\,\frac{\mathrm{kg\,m}}{\mathrm{s}^2}$.

$$\frac{\mathrm{d}N_x(x, t)}{\mathrm{d}x} = \frac{\mathrm{d}}{\mathrm{d}x}\left(EA\frac{\mathrm{d}u_x(x, t)}{\mathrm{d}x}\right), \tag{9.8}$$

where the derivative of the normal force can be replaced by the equilibrium equation
(9.5) to obtain in the general case:

$$\frac{\mathrm{d}}{\mathrm{d}x}\left(E(x)A(x)\frac{\mathrm{d}u_x(x, t)}{\mathrm{d}x}\right) = -p_x(x, t) + \frac{\gamma A}{g} \times \frac{\mathrm{d}^2 u_x(x, t)}{\mathrm{d}t^2}. \tag{9.9}$$

A common special case it obtained for $EA = $ const. and $p_x = 0$:

$$\underbrace{\frac{Eg}{\gamma}}_{a^2} \frac{\mathrm{d}^2 u(x, t)}{\mathrm{d}x^2} = \frac{\mathrm{d}^2 u(x, t)}{\mathrm{d}t^2}. \tag{9.10}$$

The analytical solution of this equation can be found, for example, in [1].

References

1. Inman DJ (2008) Engineering vibration. Pearson Education, Upper Saddle River
2. Öchsner A (2014) Elasto-plasticity of frame structure elements: modelling and simulation of rods and beams. Springer, Berlin

Printed in the United States
By Bookmasters